Advances in Intelligent Systems and Computing

173

Editor-in-Chief

Prof. Janusz Kacprzyk
Systems Research Institute
Polish Academy of Sciences
ul. Newelska 6
01-447 Warsaw
Poland
E-mail: kacprzyk@ibspan.waw.pl

For further volumes:
http://www.springer.com/series/11156

Lorna Uden, Emilio S. Corchado Rodríguez,
Juan F. De Paz Santana,
and Fernando De la Prieta (Eds.)

Workshop on Learning Technology for Education in Cloud (LTEC'12)

Editors
Lorna Uden
FCET
Staffordshire University
The Octagon, Beaconside
UK

Emilio S. Corchado Rodríguez
Department of Computing Science
and Control
University of Salamanca
Salamanca
Spain

Juan F. De Paz Santana
Department of Computing Science
and Control
University of Salamanca
Salamanca
Spain

Fernando De la Prieta
Department of Computing Science
and Control
University of Salamanca
Salamanca
Spain

ISSN 2194-5357
ISBN 978-3-642-30858-1
DOI 10.1007/978-3-642-30859-8
Springer Heidelberg New York Dordrecht London

e-ISSN 2194-5365
e-ISBN 978-3-642-30859-8

Library of Congress Control Number: 2012939197

Printed on acid-free paper

Springer is part of Springer Science+Business Media (www.springer.com)

Preface

The use of technology for learning has grown tremendously in the last decade. The need for continuous just-in-time training has made learning technology an indispensable part of life for workers. Learning technology of the future means that we can access programs anytime and anywhere. We can use our laptop or iPad or even cell phone to access educational technologies. Tomorrow's educational technologies are based on cloud computing. It's a natural extension of today's educational technologies where we access these software programs online. Cloud computing offers great potential in education by also allows students to interact and collaborate with an ever-expanding circle of their peers, regardless of geographical location.

This first workshop on Learning Technology for Education in Cloud (LTEC'12) provides opportunities for delegates to discuss the latest research in TEL (Technology Enhanced Learning) and its impacts for learners and institutions using cloud computing. Learning technology is a type of system that provides educational services to students. Nowadays, we are living in a world of increased mobility where proliferation of mobile technologies is creating a host of new anytime and anywhere contexts. The emerging social media of Web 2.0 are more flexible, sociable and more visually attractive. We live and learn in a connected world. Schools, colleges and universities must change to adapt to these new needs and expectations. This highlights the need for innovative solutions in education and learning.

In this workshop we explore a number of issues surrounding the use of technologies in learning, providing a platform for informed debate across all sectors of education and learning. The workshop allows researchers, practitioners and academics to present their research findings, works in progress, case studies and conceptual advances in areas of work where education and technology intersect. It brings together researchers across all educational sectors, from primary years, to informal learning, to higher education across a range of disciplines from humanities to computer science, media and cultural studies with different perspectives, experiences and knowledge, all in one location. It aims to help practitioners find ways of putting research into practice and for researchers to gain an understanding of real-world problems.

The LTEC2012 proceedings consists of 19 papers covering different aspects of learning technology from many different countries including, Brazil, China, Croatia, Tunisia, Taiwan, France, Slovenia, Netherlands, Spain and United Kingdom. We would to thank our program committee, reviewers and authors for their contributions. Without their efforts, there would be no workshop and proceedings.

Salamanca
July 2012

Lorna Uden
Emilio S. Corchado Rodríguez
Juan F. De Paz Santana
Fernando De la Prieta

Organization

Program Chairs

Dr. Lorna Uden (Chair)	Staffordshire University, UK
Dr. Emilio S. Corchado Rodríguez (Vice-chair)	University of Salamanca, Spain

Program Committee

Lorna Uden (Chair)	Staffordshire University, UK
Emilio S. Corchado Rodríguez (Co-chairman)	University of Salamanca, Spain
Ana Belén Gil	University of Salamanca
Adriana Berlanga	Open Universiteit, The Netherlands
D'Arcy Becker	University of Wisconsin, USA
Michael Vallance	Future University Hakodate, Japan
Mona Laroussi	National Institute Applied Science and Technology (INSAT), TUNISIA
Jane Sinclair	University of Warwick, UK
Zhiting Zhu	East China National University, China
Jules M. Pieters	University of Twente, The Netherlands
Rachid Benlamri	Lakehead University, Canada
Yu Hui Tao	National University of Kaohsiung, Taiwan
Richard	Self University of Derby, UK
Reinhard Baran	Hamburg University of Applied Sciences, Germany
Ana Borges	Instituto Politécnico de Coimbra, Portugal
François Bret	Université François Rabelais, France
Nicoletta Cocco	Ca' Foscari University, Italy
Rafael Corchuelo	University of Seville, Spain
Agostino Cortesi	Ca' Foscari University, Italy
Elisabeth Delozanne	Université Pierre et Marie Curie Paris, France
Francisco Duarte	Instituto Politécnico de Coimbra, Portugal

Richard Duro	University of La Coruña, Spain
Emanuele Pianta	FBK-irst, Italy
David Fairen-Jimenez	The University of Edinburgh, UK
Wolfgang Gerken	Hamburg University of Applied Sciences, Germany
Giacometti, Arnaud	Université François Rabelais, France
Hariharan, S.	JJ College of Engineering and Technology, India
Ivana Marenzi	L3S, Leibniz University of Hanover, Germany
Konrad Jackowski	Wroclaw University of Technology, Poland
Dominique Laurent	Cergy-Pontoise University, France
Patrick Marcel	Université François Rabelais, France
Marcus Specht	Open University of the Netherlands, The Netherlands
Antonio José Mendes	University of Coimbra, Portugal
Gabriel Michel	University Paul Verlaine - Metz, France
Leonel Morgado	University of Trás-os-Montes e Alto Douro, Portugal
Viorel Negru	West University of Timisoara, Romania
Jose Luis Nunes	Instituto Politécnico de Coimbra, Portugal
Salvatore Orlando	Ca' Foscari University, Italy
Carlos Pereira	Instituto Politécnico de Coimbra, Portugal
Pierpaolo Vittorini	University of l'Aquila, Italy
Rosella Gennari	Free University of Bozen-Bolzano, Italy
Teppo Saarenpää	Turku University of Applied Sciences, Finland
Sara Tonelli	FBK-irst, Italy
Claudio Silvestri	Ca' Foscari University, Italy
Dragan Simic	Novi Sad University, Serbia
Sorin Stratulat	University Paul Verlaine - Metz, France
Tania Di Mascio	University of l'Aquila, Italy
Zdenek Tronicek	Czech Technical University in Prague, Czech Republic
Jon Mikel Zabala	Lund University, Sweden
Luciana A.M. Zaina	University of São Carlos, Brazil

Local Organisation Committee

Juan F. De Paz Santana (Chairman)	University of Salamanca, Spain
Fernando De la Prieta (Co-chairman)	University of Salamanca, Spain
Javier Bajo	Pontifical University of Salamanca, Spain
Sara Rodríguez	University of Salamanca, Spain
Dante I. Tapia	University of Salamanca, Spain
Juan M. Corchado	University of Salamanca, Spain
Carolina Zato Domínguez	University of Salamanca, Spain
Antonio Sánchez Cabaco	Pontifical University of Salamanca, Spain

Contents

Applications

Cloud

Applications Multimedia and Games

An Approach for Supporting P2P Collaborative Communication Based on Learning Profile

Pedro F. Zanetti, Luciana A.M. Zaina, and Fábio L. Verdi

Federal University of São Carlos, Sorocaba, Rdv João Leme dos Santos,
Km110, Sorocaba, SP, Brazil
pdrowfz@gmail.com, {lzaina,verdi}@ufscar.br

Abstract. The growing use of small devices as cell phones and smartphones has requested for the development of applications in different areas. This kind of application demands for a local communication, called collaborative communication, between the devices without the Internet infrastructure, allowing the sharing and exchange of contents. In the e-learning area this is not different and these applications potentially attract the students' attention. In the meanwhile, in the e-learning area, the offering of relevant and interesting content can attract the student's attention, motivating him during the learning-teaching process. The goal of this work is to propose an approach for supporting P2P collaborative communication based on the comparison of learning profiles and learning object metadata. The learning profiles are split into dimensions based on Felder and Silverman model to attend different student preferences. A prototype was developed and tested to validate and evaluate our proposal.

1 Introduction

The miniaturization of computational devices for personal use along with recent advances in communication technologies has significantly expanded the access possibilities to a wide range of applications in several fields. Nowadays, it is possible to read e-mails, make financial transactions, share resources (hardware, software and data), access multimedia content, and enjoy a variety of other applications through a small devices.

Although the mobile devices have resources to communicate through the Internet, there are scenarios where communication is more localized, without the need of the Internet infrastructure. The daily interactions between people at work, university and where they live are much more evident when compared to global interactions. The common scenarios include collaborative peer-to-peer (P2P) interaction to share files and downloads together, interact to exchange data such as music and videos, and opportunistic communication. In these scenarios there is no need to use the infrastructure of the Internet in terms of domain name service (DNS), routing and addressing. All these aspects could and should be resolved locally through local services, allowing the collaborative communication.

The student's learning style is one of the ways to highlight the features that characterize him, such as: personal and social preferences, learning profile, and

L. Uden et al. (Eds.): Workshop on LTEC 2012, AISC 173, pp. 1–10.
springerlink.com

subject knowledge level. The exchange and sharing of materials between two students can further enrich the learning process when they use contents which are adherent to their learning profiles. The observation of learning styles provides users with different teaching strategies, meeting the student's individual needs. In this sense, it is important to highlight that the student learning style should be observed through different dimensions achieving diverse aspects of her/his preferences, such as media format and participation in group activities [4].

The dynamic linkage between contents and student's learning profile may enhance the adequacy of the learning objects that will be exchanged and shared between the students. The use of metadata standards adds quality to learning systems in the task of handling learning objects, improving their reuse and retrieval [1,2]. Learning objects are specified by fields that describe their general data (e.g., title, description, keywords), technical details (e.g., media format, size, software and hardware requirements), learning features (e.g., concrete and abstract approaches, visual and verbal elements), and other relevant metadata [3].

This work presents an approach for supporting P2P collaborative communication based on learning profile. It is proposed a model that allows small devices to share and to exchange learning objects through a P2P communication. Besides the supporting of communication technologies, it is considered the learning profile and topics of interest of the students involved in the process. Upon successful connection, the student who requested the communication will view learning objects of another device that he has interest and then he can request the transference of the learning objects. Metadata are adopted to describe the learning objects and the learner profile. An application was developed in Android platform[1] using Bluetooth technology to evaluate the proposal. Another application has developed to permit the learning objects, their respective metadata and the student's learner profile retrieval from an Amazon cloud to the device, providing a diversity of materials to the evaluation of the model. The design of learning and teaching's process is not consider into the scope of this work.

The remainder of this work is structured as follows: Section 2 explores related concepts: learning profile, learning object and collaborative communication; Section 3 presents the approach proposed here and its validation; Section 4 reports some related works; and Section 5 discusses some conclusions and outlines future works.

2 Theoretical Background

The learning profile has an important role to support the e-learning systems. Understanding the learning preferences of the student can assist in providing more meaningful objects to him. Each individual fits into a specific learning style, what makes him to adopt attitudes and behaviors that are repeated in different moments and situations [5]. Learning styles refer to highly individualized tastes and trends of a person, that influence their choices during the learning process. The student motivation can be improved when an e-learning environment supplies the student with elements that are in accordance with the individual's learning style.

There are several models used in the characterization of learning profiles, each of which is suitable for a different learning scope: the Myers-Briggs Type

[1] http://code.google.com/android/.

Indicator – MBTI, Kolb's Experiential Learning Model, the Hermann Brain Dominance Instrument (HBDI), the Honey-Mumford's Learning Styles Questionnaire (LSQ), and the Felder-Silverman Model [4]. Among the possibilities of leaning style modeling, the Felder-Silverman model [5,6] was chosen to be used in this work due to the model has the strongest emphasis on the relationship of learning styles and teaching strategies. Besides, the model is largely employed in computer and engineering courses that are important areas for us. This model uses the concept of dimensions, and therefore describes learning styles in different perspectives. The dimensions also facilitate the association of learning objects with learning profiles. Table 1 presents the model's dimensions adopted for us in this work.

Table 1 Adapted dimensions of the Felder and Silverman model.

Dimensions	Features	Learning Styles	Teaching Methods
Perception	The focus is in the best way through which the student can obtain information: contents, exercise types, for instance.	Sensing	Concrete
		Intuitive	Abstract
Presentation Format	It is related to the input. Content preferences chosen by the student such as media types.	Visual	Visual
		Verbal	Verbal
Student Participation	It represents the student preferences for the activities participation or observation.	Active	Active
		Reflective	Passive

Learning environments have different goals and of them is to offer educational material, usually called learning objects (LOs). In this context, LOs must be selected so as to correspond to the students' preferences. One of the ways to organize learning objects so that they can be used and reused systematically is through the use of descriptive metadata, that is, a set of attributes that describes learning objects. The LOM (Learning Object Metadata) standard [7] of the Institute of Electrical and Electronics Engineers (IEEE) is the most commonly metadata specification used for e-learning. The LOM standard has a structure that describes learning objects through descriptor categories. Each category has a specific purpose, such as describing general attributes of objects, and educational objectives. Table 2 shows the LOM categories adopted in this work.

Table 2 Description of LOM categories.

LOM Category	LOM Field	Characterization
General	Identifier, Type, Title, Language, Description and Keywords.	General description of the learning object.
Technical	Media Format (video type, sound), Size, Physical location, Requirements (object use: software version, for example).	Technical features description.
Educational	Interactive type (active, expositive).	Educational functions and pedagogical characteristics object description.
	Learning Resource Type (exercise, simulation, and questionnaire).	

In the communication side, the collaborative communication between two network devices can be classified into two types. The first one considers a collaborative communication when the most of the features to support collaboration occurs at local level. Functions such as routing, name service, finding context and message forwarding are performed without the use of a global network infrastructure. The second one uses a global infrastructure (Internet, for example) adopting the services provided by such infrastructure. There is also the possibility of a hybrid model that takes into account the two types of communication [8]. The collaborative communication is provided by different technologies with a variety of features such as autonomy, interoperability, available to share data, context sensitive, and the appropriate use of hardware resources.

Among the features mentioned above *context sensitivity* has become a key element in collaborative applications. A Context-Sensitive Application is able to adapt its operations without explicit intervention of the users, providing information and services that are relevant for users to perform their tasks using information taken out of the interaction context [9]. In this sense, context plays a key role to enable applications to refine the available information into relevant information, to choose appropriate actions from a list of possibilities, or to determine the optimal method of information delivery.

3 The Proposal Approach

The student's satisfaction with the materials offered by an e-learning system is fundamental in order to achieve the user approval. For the intent of offering quality, the relationship between the student and the system is built upon the tracing of personal preferences, reproducing the users' expectations. Hence, learner profile is one of the most important components of e-learning systems, storing the relevant data about user preferences [11].

Based on concepts of learner styles, learning objects and collaborative communication, this work describes a model to supply the sharing and the exchanging of learning objects through a P2P communication, concerning in the student's learning profile. It was shared learning objects to the student involved in the communication, organised according to the LOM standard, based on the observation of the following criteria: keywords describing the students' interests and students' learning profiles. The next subsections will report the elements of our proposal.

3.1 Requirements

For the proposed approach be adopted, some requirements must be attended. The first one is the learning objects available on the collaborative communication must be cataloged using LOM standard. The proposal adopts only the LOM categories and fields described in Table 2.

Another requirement is that the learner model needs to report the student learning profile using dimensions based on Felder and Silverman (Table1). A learner model contains relevant information to attend a system needs to automate tasks. The learner model adopted in this work contains the student identification, keywords that represents his learning interests and learning profile split into three dimensions: perception, presentation format and student participation. A local repository of learning objects, their respective metadata and the student's profile must be stored into the devices. This repository can be updated when the student's searches changes.

3.2 The Collaborative Communication

In this paper it was considered the collaborative communication with local support of network functions without the use of Internet infrastructure. The devices involved in a P2P communication can perform both roles in different situations: client and server. The roles are dependent on who starts the communication requesting that is the client. The interaction starts after two devices have been found and recognized. When a device's Bluetooth system is active it scans for other active devices. The student selects the device he wishes to connect and then the requesting device becomes the client and the other the server. As the result, the communication is established and the collaborative process begins automatically. The scan and the connection process can be developed according to the technology adopted in the communication. The different technologies supply the developers with API to provide these implementations.

The proposed model contains two main components ***Context Manager*** and ***Object Sharing Manager***, which both are composed of other subcomponents as shown in Figure 1.

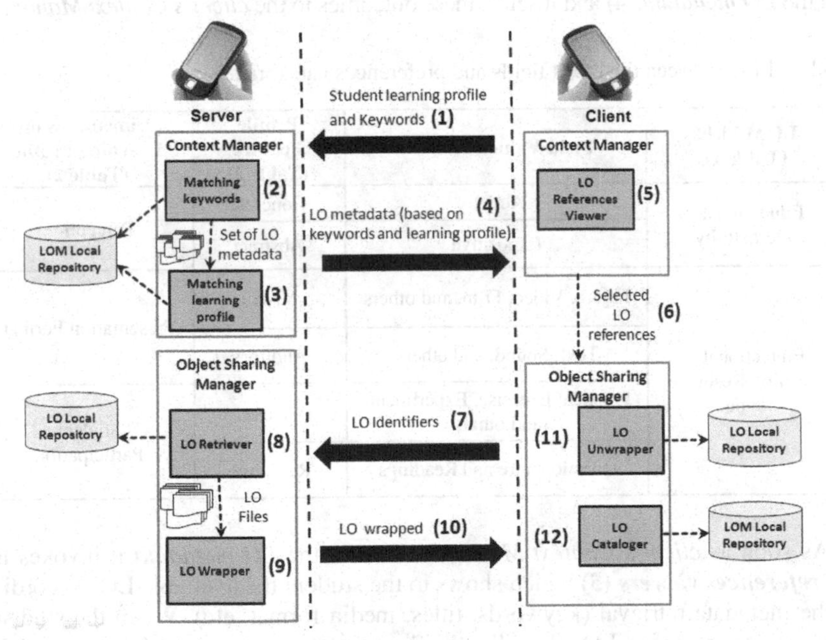

Fig. 1 P2P collaborative communication proposal.

Context Manager is responsible for providing the communication and the interaction between both P2P elements of the application: client and server. In the server side it deals with the matching and retrieval procedures whilst in the client side it presents the learning objects retrieved from the server. ***Object Sharing Manager*** manages the object sharing between applications and devices, wrap and unwrap objects and save and request objects. To enable the collaborative communication the ***Context Manager*** must be running on both devices: the client and the server.

Once communication is set out the *client's Context Manager* sends to the **server's Context Manager** a file with the *student's learning profile* (split in dimensions) and the *keywords* of student interests **(1)**. The *server's Context Manager* invokes the ***Matching Keywords*** **(2)** (subcomponent) which looks for *learning objects metadata* in the *LOM local repository*. The method seeks for learning objects which the metadata fields (*title*, *description* and *keywords)* match with the keywords send by the client. The method returns a set of ***learning object metadata (LO metadata)*** that fits with the *keywords*.

Considering only the learning objects' references got in the previous step, the *server's Context Manager* performs the next step based on the criterion *learning profile*. In the ***matching learning profile*** **(3)**, the dimensions of the student's learning profile (*Perception, Presentation Format and Student Participation*) are compared to the *Interactive* and *Learning Resources* fields of the *Educational category of LOM standard* (Table 1). Table 3 presents the binding between the fields of the LOM standard (describing the learning content – Table 1) and the students' Preference Categories (Felder-Silverman – Table 2). The method returns to the **server's Context Manager** a set of metadata that fits the dimensions of the *client's learning profile* and the ***LO metadata*** **(4)** and it sends these outcomes to the *client's Context Manager.*

Table 3 Link between the LOM fields and preferences categories.

LOM Field (Table 1)	Field Values(Table 1)	Profile Feature (Table 2)	Dimensions of Learning Profile (Table 2)
Educational - Interactivity	Active	Concrete	Perception
	Expositive	Abstract	
Educational - Learning Resource Type	Figure, Video, Film, and others	Visual	Presentation Format
	Text, Sound, and others	Auditory	
	Practical Exercise, Experiment, and others	Active	Student Participation
	Questionnaire and Readings	Reflexive	

As soon as *client's Context Manager* receives the *LO metadata* it invokes the ***LO references viewers*** **(5)** which shows to the student the available LOs according to the metadata retrieval (keywords, titles, media format, etc). When the students chooses one or more LO, the *client's Context Manager* sends the *selected LO references* to the **client's *Object Sharing Manager* (6).**

The *server's Object Sharing Manager* receives from the client the ***LO Identifiers*** **(7)**. The ***LO retriever*** retrieves the respective LO in the *LO local repository* **(8)**. After this, the *LO files* are wrapped by the ***LO wrapper*** **(9)** and sent by *server's Object Sharing Manager* **(10)** to the client. In the client, *LO files* are unwrapped and stored in a *LO local repository* and then cataloged the in a *LOM local repository*, respectively performed by ***LO Unwrapper*** **(11)** and ***LO Cataloger*** **(12)**. The client can provide the new LOs in a future collaborative communication.

3.3 The Validation's Approach

In order to supply the proposal approach in the validation process we developed a mobile application that allows users to download the learning objects, their respective metadata and the student's learner model previously registered from a cloud application, called Web Collaborative Learning (WCL). The student's learner model is composed of the keywords (stating his learning preferences) and his learning profile spitted into dimensions. By using the WCL, the user may look for learning objects in the Web, select ones that he has interests, and register the value of LOM field creating the LO metadata. The user's learning profile was extracted from the outcomes of the questionnaire of Soloman e Felder [11] that the user answered in his first access to WCL.

WCL application is hosted in an Amazon[2] cloud and uses the SQLite database 3.6.20[3]. Our mobile module connects to the cloud database and before the objects be downloaded to the mobile device it shows the description of the learning objects catalogued (LOM fields as title, media format, etc). The user chooses the learning objects and the mobile module performs the download, storing the objects and their metadata in a local repository. The user can also download from the cloud his learner model. After that, the user does not need the Internet infrastructure to connect and to share the learning objects with other devices.

We have implemented a prototype application of e-learning P2P in an Android platform based on the proposal model, simulating the communication in an emulator previously. After the first tests, we have run the application in two devices which support of Bluetooth API: Samsung Galaxy 5 (Android 2.1) and Motorola Milestone (Android 2.0). We used the Bluetooth API for implementing the functions of communication as forwarding messages, service names and routing.

Before starting the communication with the devices, we downloaded from the Amazon cloud a set of ten learning objects and their metadata to the Galaxy 5 and a set of seven learning objects to the Motorola. The learner model of the users is also downloaded. The learner model saved in Galaxy has the keywords *computer networking, mobile, cloud* and *android* and its *perception, presentation format* and *participation* dimensions have the follow values: *sensing, visual and active*.

As an example, we report an illustrative operation in which Galaxy is the client and Motorola is the server (Figure 2 shows the snapshots): **(a)** Galaxy finds Motorola and starts the communication; ***Context Manager of Motorola*** receives the

[2] http://aws.amazon.com/pt/

[3] http://www.sqlite.org/

keywords and the *learning profile* from ***Context Manager of Galaxy***; **(b)** after the matches process in the ***Context Manager of Motorola***, ***Context Manager of Galaxy*** receives *four* of seven available LOs that match to its learner model (*keywords* and *learning profile*), showing the options to the student. By selecting the *cloud computing* learning object, the ***Context Manager of Galaxy*** invokes its ***Object Sharing Manager*** that requests to the server the transference of the learning object; **(c)** the ***Object Sharing Manager of Motorola*** retrieves and wraps the *cloud computing* learning object, delivering it to the client. The ***Object Sharing Manager of Galaxy*** saves the object and its metadata, forwarding the object reference to its ***Context Manager*** that presents the learning object to the student. In Figure 2 we can observe the adhesion between the presentation format of the learning object (a picture) and the value of the student presentation format dimension (visual).

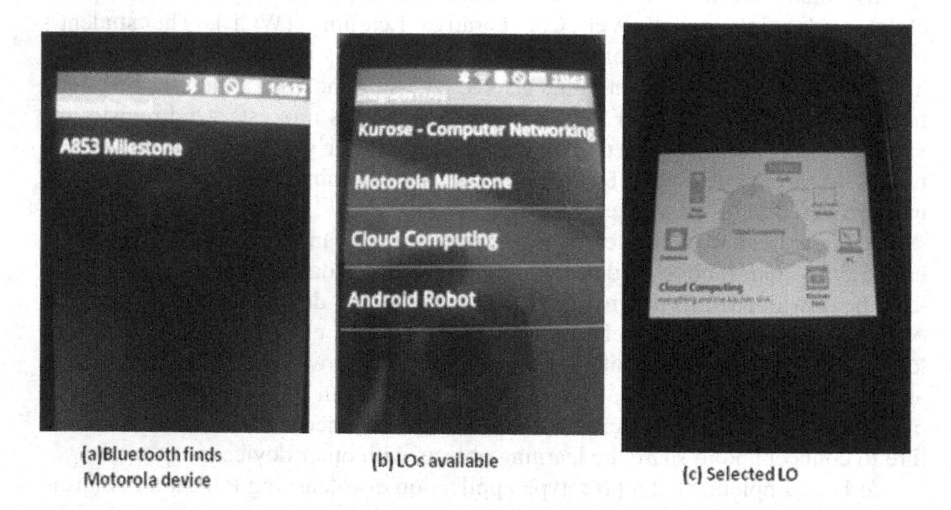

Fig. 2 Screen pictures of proposed approach in operation.

4 Related Works

There are e-learning application examples found in the literature that works with learning profile, learning objects and P2P communication. We focused in the follows which have relationship with our approach.

Felder and Silverman [5] proposed a model based on dimensions of learning and teaching styles, creating a relationship between learning styles and teaching strategies that can be used to support the students' learning styles. The authors argue that the learning style should be observed by four different behaviour dimensions. The model of Felder and Silverman aggregates a wider set of desirable features: simplicity, open and wide use, and online availability. Nonetheless, it has not proposed how to realize a dynamic linkage between the content and the profile.

Milošević et al. [12] proposed the adoption of a learning style that allows the system to build learning workplaces, bounding learning content and learning styles through the SCORM (Sharable Content Object Reference Model) [2]. Although this proposal adopts a standard to support the concept specification, it does not consider the learning profile according to dimensions.

EduSHARE [13] e-learning application proposes an e-learning application that allows the data sharing in a P2P communication between student and teacher, and student and student. The main feature is to support the feedback on understanding of what is being discussed in the class through questions sending to the student by the teacher. The student response may be sent to the other students sharing the impressions and doubts. Nevertheless, the proposal does not adopt standards to organize the data and learning profile to offer contents to the students.

Honey and Mumford's learning styles questionnaire was employed by Lowery [14]. The assessment phase was conducted with the offering of activities, and with the identification of students' styles. The author reports the problems in assisting some styles in disciplines with practical nature, bringing new challenges to future online lectures planning.

5 Conclusions and Further Works

In this work, we have raised up an approach that supplies the sharing and the exchanging of learning objects through a P2P collaborative communication, without usage of Internet infrastructure. The focus concerns in the student's learning profile and its linkage to learning objects for automatic content offering. To do so, we use the Felder-Silverman Learning Style Model along with the IEEE LOM standard, a combination proposal that, extending former works, can suitably relate learner profiles and learning objects, automatically, in different fields of learning, and consistently reflecting the intrinsic style of the students. We use a multiple-matching to select the learning objects according to two criteria: keywords and profile (style).

A prototype of our proposal was developed and experimented in two different mobile devices through Android platform. A mobile module, which connected to a cloud application database, was developed to aid users to download the proposal requirements (learning objects, metadata and learner model) data. This practice improved our experiment, allowing us to focus in our approach results. Another important point is that the cloud application (WCL) has allowed the students to catalogue various learning objects they have interest. However, it is not download into the device all the catalogued learning objects for reasons of storage space use.

Further works include adding new features to the matching process to consider other context elements beyond of learning profile. Such new features include student physical localization and device resources (memory, processor, screen size etc), features that must be considered when choosing the most adequate content. We are also interested in investigating the usage of different protocols to tranfer data such as protocol buffers[4] and thrift [5]. These protocols will enable the optimization of the exchange of data between the cloud and the devices and between the devices.

---

[4] http://code.google.com/p/protobuf/

[5] http://thrift.apache.org/

Acknowledgments. We thanks CNPq and FAPESP (Brazil) for financial support.

References

1. Devedžic, V., Jovanovic, J., Gaševic, D.: The pragmatics of current e-learning standards. IEEE Internet Computing 11(3), 19–27 (2007)
2. Devedžić, V., Gašević, D., Djurić, D.: Clarifying the meta. International Journal of Information and Communication Technology 1(2), 148–158 (2008)
3. Zaina, L.A.M., Bressan, G.: Learning objects retrieval from contextual analysis of user preferences to enhance e-learning personalization. In: Proc. of IADIS International Conference WWW/Internet 2009, pp. 237–244 (2009)
4. Zaina, L.A.M., Rodrigues Jr., J.F., Cardieri, M.A.A.C., Bressan, G.: Adaptive learning in the educational e-LORS system: an approach based on preference categories. International Journal of Learning Technology 6(4), 341–361 (2011)
5. Felder, R.M., Brent, R.: Understanding Student Differences. Journal of Engineering Education 94(1), 57–72 (2005), doi: 10.1.1.133.171
6. Felder, R.M., Silverman, L.K.: Learning and Teaching Styles in Engineering Education. Journal of Engineering Education 78(7), 674–681 (1988), doi: 10.1.1.92.774
7. IEEE LOM, Draft standard for learning object metadata, `http://ltsc.ieee.org/wg12/index.html` (accessed June 25, 2009)
8. Jung, S., Lee, U., Chang, A., Cho, D., Gerla, M.: BlueTorrent: Cooperative Content Sharing for Bluetooth Users. In: IEEE PerCom 2007, White Plains, NY, March 19-23 (2007)
9. Vieira, V., Tedesco, P., Salgado, A.C.: Designing context-sensitive systems: An integrated approach. Expert Syst. Appl. 38(2), 1119–1138 (2011)
10. Cramer, H., Evers, V., Ramlal, S., Someren, M., Rutledge, L., Stash, N., Aroyo, L., Wielinga, B.: The effects of transparency on trust in and acceptance of a content-based art recommender. User Modeling and User-Adapted Interaction 18(5), 455–496 (2008)
11. Soloman, B.A., Felder, R.M.: Index of Learning Styles Questionnaire (1997), `http://www.engr.ncsu.edu/learningstyles/ilsweb.html` (accessed June 19, 2008)
12. Milošević, D., Brković, M., Debevc, M., Krneta, R.: Adaptive Learning by Using SCOs Metadata. Interdisciplinary Journal of Knowledge and Learning Objects 3(1), 163–174 (2007)
13. Angelaccio, M., Buttarazzi, B.: Adaptatative Peer to Peer Data Sharing for Technology Enhanced Learning. In: Lytras, M.D., Ordonez De Pablos, P., Avison, D., Sipior, J., Jin, Q., Leal, W., Uden, L., Thomas, M., Cervai, S., Horner, D. (eds.) ECH-EDUCATION 2010. CCIS, vol. 73, pp. 425–430. Springer, Heidelberg (2010)
14. Lowery, C.: Adapting to student learning styles in a first year electrical/electronic engineering degree module. Journal of the Higher Education Academy Engineering Subject Centre 4, 52–60 (2009)

CAFCLA: A Conceptual Framework to Develop Collaborative Context-Aware Learning Activities

Óscar García, Ricardo S. Alonso, Dante I. Tapia, Elena García, Fernando De la Prieta, and Ana de Luis

Department of Computer Science and Automation, University of Salamanca. Plaza de la Merced, s/n, 37008, Salamanca, Spain
{oscgar,ralorin,dantetapia,elegar,fer,adeluis}@usal.es

Abstract. Advances appeared in Information and Communication Technologies along last years have given raise to new interaction ways between people and technology. Ambient Intelligences (AmI) is a multidisciplinary research area which promotes the use or technology in a transparent way to facilitate everyday tasks. Education is one field that benefits from AmI: collaboration between students in innovate ways, data acquisition from the context or real time location systems enhance learning processes. This paper presents CAFCLA, a framework aimed at designing, developing and deploying AmI-based educational scenarios where collaboration between students and contextual information are available every time and everywhere through multiple resources and communication protocols.

Keywords: Ambient Intelligence, Computer Supported Collaborative Learning, Context-aware Learning, Wireless Technologies, Real Time Locating Systems.

1 Introduction

In recent years there has been a technological explosion that has flooded our society with multiple and different technological devices. Similarly, devices improve their processing and storage capacity and storage, their user interfaces or their communication skills day by day. Thanks to these advances, we are currently surrounded by technology that has changed our habits and customs [1]. All this has given cause to appear fields such as Ambient Intelligence, whose main objective is to simplify the use of technology to improve the quality of life of users [2].

Education is one of the areas in which Ambient Intelligence presents a greater potential as it provides new ways of interaction and communication between individuals and technology systems [3]. The usage of Communication and Information Technologies (ICT) has been present in educational innovations over recent years [4], modernizing the traditional transmission of content through electronic

L. Uden et al. (Eds.): Workshop on LTEC 2012, AISC 173, pp. 11–21.
springerlink.com

presentations, email or more complex learning platforms such as Moodle[1] or LAMS[2] and fostering collaboration between students (Collaborative Learning) [5]. Besides the use of those general-purpose tools in education, other tools that make more concrete use of technology have appeared. This applies to those that make use of Context-awareness information and ubiquitous computing and communication, fundamental parts of Ambient Intelligence [6].

The inclusion of Context-awareness in educational environments and processes refers to *Context-aware Learning* [7], a particular area of application of *Context-aware Computing* [8]. Moreover, being able to know, characterize and customize the context that surrounds a leaning situation at a time and place allows flexibility in the education process, so learning does not only occurs in classrooms, but in a museum, park or any other place [9], obtaining ubiquitous learning spaces. Thus, there is an extensive literature that addresses the problem of this kind of learning, highlighting those works that attempt to solve contextual information acquisition and providing data to users [10, 11, 12]. The use and integration of different technologies and the approach to specific learning activities characterize analyzed solutions. However, there is a lack of solutions that attempt to facilitate the work of educators. The complexity of understanding and use of the technology and solutions in the aforementioned works does not allow a wide use of them.

This paper presents a conceptual framework aimed at designing, developing and deploying AmI-based educational scenarios. Educators are able to characterize the context where the learning activity will occur through the creation of a *world model* in which locate data collectors (e.g. sensors), indentify and characterize areas of interest (e.g. paintings in a museum), etc. Moreover, the collaboration between students and the customization of the information available is also provided and can be integrated in the activity design.

The following section describes the background and problem description related to the approach presented. Then, the main characteristics of the framework proposed are described: what kind of activities is covered, how educators can create an activity, what technologies are used, which parts compose the framework and how they work together. Finally, the conclusions and future work are depicted.

2 Background and Problem Description

A growing interest in educational software, commonly known as e-learning, has appeared over recent years [13]. Among the wide range of existing educational software are CSCL (*Computer Supported Collaborative Learning*) applications [14]. A collaborative learning system consists of a set of tools that facilitate the implementation, development and deployment of learning activities. Those activities allow different ways of interaction between participants involved that activate learning mechanisms [15]. CSCL has become an important research field within

[1] http://moodle.org
[2] http://lamsfoundation.org/

education that attracts different interests from the purely educational to those focused on improving human-computer interaction [5].

On the other hand, contextual information includes any data that can be used to characterize a person, place or object that is considered relevant to the interaction between users, user and applications or systems or between systems or applications [16]. In addition to the relevant information that context provides, it is important to consider other parameters that relevantly affect this type of information, such as identification, time and location [6]. The information exchange taking place between technology and users, in order to contextualize an environment in which learning takes place, and customize the content of the learning activity can be understood as collaboration. Thus, Context-aware Learning must take into account the interactions between people and the different technological components of the system in all its combinations.

2.1 Providing Context-Aware to Learning

Providing contextual information and fostering collaboration between students benefit the learning process [3]. Moreover the combination of Collaborative and Context-aware Learning naturally leads to think about ubiquitous learning spaces, characterized by "*providing intuitive ways for identifying right collaborators, right contents and right services in the right place at the right time based on learners surrounding context such as where and when the learners are (time and space), what the learning resources and services available for the learners, and who are the learning collaborators that match the learners' needs*" [17].

A better understanding of environment through technology allows educators to customize the content provided to students. Similarly, technology facilitates the interaction with the environment and between students. This should try to be reached in a way as transparent and ubiquitous as possible. The technologies used for the collection of contextual information and for the communication between different devices are the cornerstone of the different works presented here. Literature about Context-aware Learning proposals has been deeply reviewed in this work. Some of the most representative works are classified in this paper, following technological criteria related to communications and data collection.

A first approach to provide contextual information is "tagging the context". RFID (Radio Frequency IDentification) is the most spread technology [18]. While other technologies such as NFC (Near Field Communication) [19] or QR Codes (Quick Response Codes) [20] are growing fast. As can be seen in the usage of Active RFID, location and context-awareness are closely related: knowing precisely location of objects and people allow determine what is surrounding them and, consequently, characterize the context they are involved. GPS (Global Positioning System) is the most used technology to provide location in Context-aware Learning [20, 21, 22]. This location system provides a high accuracy level and is currently implemented in a wide range of smart phones and mobile devices. In those cases, the mobile device provides a position to the system. Those solutions are used in different scenarios as planning routes [22] or manage a student scheduler [21]. However, most of those works do not implement a concrete use case, but propose a general purpose model in which GPS technology is included to facilitate the provision of contextual data.

Moreover, GPS technology does not work indoors because of the direct vision necessary between satellites and devices. Indoor environments are very common in learning: museums, laboratories or the school are places where activities that requires mobility can be developed. Trying to cover this lack location systems that user Active RFID [18] or Wi-Fi [12] are used. Both cases the performance of systems is similar: student's position y determined by the access point is giving coverage in this moment. This type of approach has significant limitations when developing context-aware learning activities: location accuracy is poor. This situation presents an important problem when areas where context information is different are close (e.g. two paintings in a museum).

Changing the way of contextualize the environment where the learning activity occurs, some tendencies use sensor networks [12]. A sensor is an electrical or mechanic device that measures a physical magnitude. In this case environmental characterizations is reduced to those data provided by the sensors. In order to make the learning process more transparent to students, ad-hoc networks are considered to collect and transport data from sensors to remote points [11]. This networks facilitate the connection between devices anytime and anywhere without a previous infrastructure.

2.2 Problem Description

The review of the literature evidences some lacks in the Context-aware Learning systems proposed until now. Even some works try to combine different technologies to cover as much situations as possible [12], most of them only cover specific learning situations, as those where tagging context with RFID/NFC [19] is necessary or those where learning occurs outdoors [22]. The combination of both situations is only addressed by M2learn [12]. However, this solution does not provide a precise and efficient Real Time Locating System or the possibility to integrate wireless sensor networks, excepts RFID systems.

On the other hand, none of the solutions mentioned before takes into account Ambient Intelligence issues. The proposed solutions focus their work on the architectural description, framework integrators or end-user applications whose designs have not taken into account how complex will be them for educators or students. Some aspects such as designing intuitive and attractive interfaces or abstracting end users from the complexity of technology, issues on which Ambient Intelligence pays special attention, are not taken into account. Thus, if these aspects are excluded from the solution design process, the final result may be rejected by students and educators.

For this reason, the design process must take into account, from the beginning, the opinion of all the stakeholders [5], educators, designers and developers. This way is easier to accomplish with Ambient Intelligent issues related to user interfaces and usability of final applications.

Moreover, works analyzed in this review do not include mechanisms for data or communication management. Ambient Intelligence emphasizes the transparency of technologies for users. In addition, technology is used to ease ordinary task or improve activities and quality of life. In this sense, systems that combine different technologies do not facilitate mechanisms to change between them (e.g. different communication protocols) attending to the needs of a situation. Similarly, data have to be managed in an intelligent and efficient way. Most of the literature reviewed does not include this issue, using only standard data repositories that only consider persistency and consistency [29]. Functionalities like data redundancy to solve network failures helps make the system dynamic and benefits data accessibility with independence of the place and the moment.

Collaboration between students is another issue not considered by many proposals [11]. It is well known that collaboration benefits the learning process [3]. Including mobile devices and wireless communication protocols in any learning design that requires mobility (as discussed in this paper) is nowadays necessary. Mobile devices easily connect each other so including collaboration between students is easy, increasing the variety of activities and improving learning process.

Furthermore, this work is developed following Ambient Intelligence issues. Some of those issues, as personalization of context provided or transparency and ease of use for educators and users are vaguely taken into account. Moreover, the inclusion of reasoning mechanisms facilitate the personalization of data provision or the communication management of this kind of complex systems [23]. The present work considers all these issues when describing and designing the framework presented next.

3 CAFCLA: Context-Aware Framework for Collaborative Learning Applications

CAFCLA (*Context-Aware Framework for Collaborative Learning Applications*) is a framework aimed at designing, developing and deploying AmI-based educational scenarios, focusing in collaborative and context-aware activities. The framework integrates a set of wireless context-aware technologies (e.g. GPS, ZigBee, Wi-Fi, or GPRS/UMTS). Those technologies allow establishing collaborative activities based on Ambient Intelligence among students and teachers. In this sense, communication models vary dynamically depending on the activity; for example, following a client-server model to perform a data query or forming an ad-hoc network to gather contextual information. Thus, the contextual information is always available and may be modified every time.

3.1 Main Characteristics

As mentioned before, the biggest challenge of CAFCLA is to ease the design, develop and deployment of Ambient Intelligences educational applications in which contextual data and collaboration are required. This implies that educators are able to choose what, where and how they want to develop a collaborative learning activity using contextual data. The main resources that CAFCLA offers to educators in order to reach these objectives are:

- **Identification:** each participant or object that takes part in the learning activity is uniquely identified. This way the learning process can be customize for each student: once the environment is characterize, educator can choose what information is given to each student, filtering data in order to the student's identification. Moreover, educators can create profiles for groups of students in order to create group activities or manage information in an easy way.
- **Wireless Sensors Network:** educators can collect environmental data through sensors. CAFCLA integrates n-Core® ZigBee wireless sensor network [24]. n-Core® platforms consists of a set of wireless devices with reduced energy consumption and physical size. Each of them collects one or more physical magnitude used to characterize the environment. They use ZigBee communication.
- **GPS Localization:** to contextualize environments a provide users' position outdoors, CAFCLA integrates GPS localization.
- **ZigBee Real Time Location System:** to solve the inability to provide indoor location of GPS, CAFCLA integrates n-Core® ZigBee Real Time Location System [24]. The deployment of this systems consists of a set of beacons strategically located along the area that needs localization. Its accuracy, under 1 meter, allows educators to define as many areas of interest as they desire in which customize contents of includes environmental information.
- **Collaboration:** thanks to the different communication protocols implements by the used devices. students are able to collaborate between them through different activities proposed by educators. Moreover, the uniquely identification of each student can be used by educator to decide who collaborate with whom, so they can control at any moment the develop of the activity.
- **Disconnected operation mode:** a failure in the connection to the Internet can avoid data or activity access and, consequently, stop the activity. As ZigBee protocol is integrated in CAFCLA, devices are able to form ad-hoc networks to communicate each other (as sensors do) without the requirement of a previous infrastructure. Students can collaborate each other using this kind of networks.
- **Analysis interaction:** ad-hoc networks, outdoors and indoors location and uniquely identification of each student provides interesting information about interactions between students that can be analyzed in order to evaluate and improve the learning process.

This set of resources allows that educators only think in the learning process, abstracting the technological layer and the complications and difficulties that this introduces. CAFCLA provides educators an application in which graphically design the educational scenario where the activity is taking place. Once collected the basic data, the framework tells the educator how many and what kind of devices are needed, where each of them should be located, what services are necessary to use and how and where insert contextual data into the system. On the other hand, CAFCLA provides developers with programming functions that they have to use to develop the application designed. Thus, developers can abstract from any problems of logical, devices, configurations, etc.

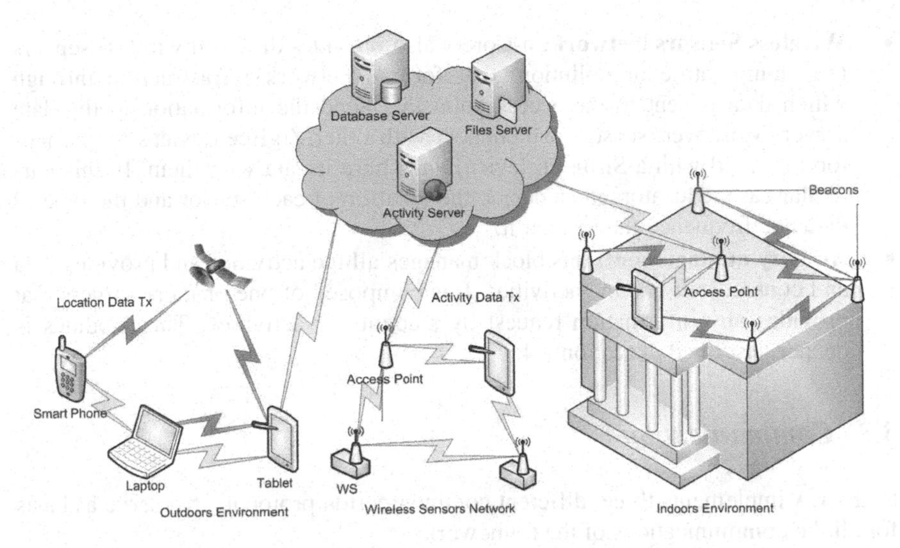

Fig. 1 Basic architecture of CAFCLA.

3.2 Architecture

As can be seen in Figure 1, three are basic cases for using CAFCLA: learning outdoors, learning indoors and learning using wireless sensor networks. All of them can be used at the same time to support the same application. Moreover, data management is located in one or more servers. The physical components of each case and it basic performance are as follows:

- **Outdoors environment:** this functionality requires, at least, a GPS provided device. Moreover a maps platform (e.g. Google Maps) is needed. When designing activities, educators draw a concrete area in the map. Then includes al the contextual information related to that area, including different versions of the information to be used in different activities. Then the system associate an area to one or more descriptions. Then, when students are developing the activity GPS embedded in the device is continuously transmitting its position. When the student enters into an area he receives contextual information accordingly to the design of the activity.
- **Indoors environment:** n-Core® platform ease the location process. The zone where location system is implemented is provided with a set of beacons (Sirius D). Those beacons are able to communicate each other and send information about the location of a student to the access point, Each student is provided with a ZigBee device (Sirius B) that communicate with its closer beacons that gather different data and send it to the access point. The access point send all the data information to the activity server where a location engine calculate the position of the student. The way contextual information is included in the system is identical to the previous case.

- **Wireless Sensors Network:** n-Core® also provides different wireless sensors (e.g. temperature or pollution) that form a network infrastructure through which data is sent to the access point that send the information to the data server. Moreover, sensors can connect with other ZigBee devices (e.g. a laptop provided with a Sirius B device) and share it data with them. In this particular case, educators must decide the location of each sensor and the type of data and frequency they gather it.
- **Activity management:** this block manages all the activities and provides data and content to users and activities. It is composed of one or more servers that provides any information request by students or activities. This modules is deeper described in section 3.4.

3.3 Communication

CAFCLA implements three different communication protocols that serve as basis for all the communications of the framework:

- **Wi-Fi:** most of mobile devices, as laptops, tablets or smart phones are provided with Wi-Fi communication. Moreover, Wi-Fi internet access is provided in great part of the placer where daily life takes place, such as homes, schools, universities, coffee shops, museums even public spaces as parks or places. thus, it is easy to access to remote resources at any time.
- **GPRS / UMTS:** CAFCLA provides the possibility to use 3G communications to access the resources of the learning activity. For these places where a Wi-Fi Internet connections is not available, if the student's device has 3G connection, he is able to access to any resource.
- **ZigBee:** besides the connections provided by Wi-Fi and GPRS capabilities to access to remote resources, CAFCLA also includes the possibility to integrate ZigBee communication protocol. This is helpful when deploying ad-hoc networks that foster peer-to-peer collaboration between students without the needed of having Internet access or when a device gather data from different sensors.

CAFCLA integrates these protocols so that the communication between devices and people is transparent both to the design activity (taking as constraints only the functionalities of the framework and not technological ones) as to when carrying out specific activities.

In addition to integration of these communication protocols in the framework, the biggest challenge of CAFCLA is to manage them in an intelligence and effective way. Thus, the network management system hides the complexity of decide which are the available networks, select the most appropriate network depending of the activity and the availability and other tasks that must be transparent to educators and students, such as changing the communication protocol when necessary. He is responsible to detect at any time which communication networks are available, choose the most appropriate to develop the activity, offer and provide peer communication (e.g. sensor-device or two students to collaborate).

3.4 Data Management

Once defined all the communications protocols involved in the framework it is necessary to introduce how information and activities are managed. In this sense, management has been divided in three main parts:

- **Activity Server:** this server is responsible for maintaining a general repository with all the activities and services developed using CAFCLA. Thus, all of them are reusable by other people.
- **Database Server:** provides all the logical information of the system and subsystems developed. It has a structure of great directory that stores and receive queries such as services used in activities, customized information, contextualization of environments, location or sensor networks available, etc.
- **Files Server:** data is completely separated from logical structure. A file repository have been designed in order to maintain integrity, security, redundancy and availability of the data.

Access to this information is marked by the pattern followed to develop the activity designed. A simple reception of contextual information accesses the different servers as follows: one access point transmits the data necessary to calculate the position of a student to the Activity Server, more specifically the service capable of calculate its position. Once the position is calculated and passed to the manager of the activity (responsible for maintaining a logical flow of it), it requests the data context for this position, querying the Database Server. This server will refer to specific data that may be obtained from the File Server.

On the other hand, systems designed using CAFCLA must be dynamic and must work sometimes without an Internet access. For this reason a data management system is designed and developed. It is responsible to decide which data should be stored in each device depending of the location where the activity is taking place and the resources there available.

4 Conclusions and Future Work

Communication and Information Technologies are in constant rise from several years ago. Different devices and forms of communication have appeared and have been adopted in daily life naturally. However, much work is necessary to manage and coordinate all the elements involved until raise an end-user solution.

Education is a field where technology use is widespread. The use of new mobile devices, context-aware technologies or communication protocols promote the emergence of new learning scenarios. More specifically, location technologies (e.g. GPS) or wireless communication protocols (e.g. Wi-Fi or GPRS) are included in a large number of proposed solutions. However, they are usually focused to cover concrete scenarios, so it is difficult to extrapolate solutions from a scenario to another.

Moreover, the design of aforementioned solutions are usually made from a highly technical point of view. A global understanding of the learning process requires the participation of all the stakeholders, educators, designers and developers. Collaboration among them is an essential part for the success of the proposal. Furthermore, all stakeholders participation in the whole process benefits the inclusion of the solution in a wider range of learning scenarios.

Then, it is necessary to design and develop a set of tools that provides a basis to easily design, develop and deploy the learning activities this works refers. This paper presents CAFCLA as an example of this kind of tools. CAFCLA is a framework that integrates different context-aware technologies, real time location systems and communication protocols to abstract educators and developers of learning activities on the complexity of the simultaneous use of different technologies involved. In this case, CAFCALA is focused on facilitating the design and develop of collaborative context-aware learning activities.Future work includes the inclusion of management systems in the framework. Two will be necessary: first, a communication management system that will provides each moment the most appropriate communication protocol to be used. In second place, a data management system will be also integrated to improve the availability of resources and anticipate to situations in which general purpose resources are not accessible. By the moment, Multi-agent Systems is the technology that has considered more suitable. However, a deep analysis and comparison will be done in order to objectively determinate the technology that will be used.

Acknowledgments. This project has been supported by the Spanish Ministry of Science and Innovation (Subprograma Torres Quevedo).

References

1. Jorrín-Abellán, I.M., Stake, R.E.: Does Ubiquitous Learning Call for Ubiquitous Forms of Formal Evaluation?: An Evaluand oriented Responsive Evaluation Model. Ubiquitous Learning: An International Journal 1 (2009)
2. Corchado, J.M., Bajo, J., de Paz, Y., Tapia, D.I.: Intelligent environment for monitoring Alzheimer patients, agent technology for health care. Decision Support Systems 44(2), 382–396 (2008)
3. García, Ó., Tapia, D.I., Alonso, R.S., Rodríguez, S., Corchado, J.M.: Ambient intelligence and collaborative e-learning: a new definition model. Journal of Ambient Intelligence and Humanized Computing, 1–9 (2011)
4. Scardamalia, M., Bereiter, C., McLean, R.S., Swallow, J., Woodruff, E.: Computer-Supported Intentional Learning Environments. Journal of Educational Computing Research 5(1), 51–68 (1989)
5. Gómez-Sánchez, E., Bote-Lorenzo, M.L., Jorrín-Abellán, I.M., Vega-Gorgojo, G., Asensio-Pérez, J.I., Dimitriadis, Y.A.: Conceptual framework for design, technological support and evaluation of collaborative learning. International Journal of Engineering Education 25(3), 557–568 (2009)
6. Traynor, D., Xie, E., Curran, K.: Context-Awareness in Ambient Intelli-gence. International Journal of Ambient Computing and Intelligence 2(1), 13–23 (2010)
7. Laine, T.H., Joy, M.S.: Survey on Context-Aware Pervasive Learning Environments. International Journal of Interactive Mobile Technologies 3(1), 70–76 (2009)
8. Dey, A.K.: Understanding and Using Context. Personal and Ubiquitous Computing 5(1), 4–7 (2001)

9. Bruce, B.C.: Ubiquitous learning, ubiquitous computing, and lived experience. In: Cope, W., Kalantzis, M. (eds.) Ubiquitous Learning, pp. 21–30. University of Illinois Press, Champaign (2008)
10. Abowd, G.D., Atkenson, C.G., Hong, J., Long, S., Kooper, R., Pinkerton, M.: Cyberguide: A mobile context-aware tour guide. Wireless Networks 3(5), 421–433 (2006)
11. Chen, T.-S., Yu, G.-J., Chen, H.-J.: A framework of mobile context management for supporting context-aware environments in mobile ad hoc networks. In: Proceedings of the 2007 International Conference on Wireless Communications and Mobile Computing, pp. 647–652 (2007)
12. Martín, S., Peire, J., Castro, M.: M2Learn: Towards a homogeneous vision of advanced mobile learning development. In: 2010 IEEE Education Engineering (EDUCON), pp. 569–574 (2010)
13. Clark, R.C., Mayer, R.E.: E-Learning and the Science of Instruction: Proven Guidelines for Consumers and Designers of Multimedia Learning. John Wiley & Sons (2011)
14. Dillenbourg, P.: What do you mean by Collaborative Learning? Collaborative Learning. In: Cognitive and Computational Approaches, pp. 1–19. Elsevier Science Ltd, Oxford (1999)
15. Koschmann, T.: CSCL: Theory and practice of an emerging paradigm. Lawrence Erlbaum, Mahwah (1996)
16. Abowd, G.D., Dey, A.K., Brown, P.J., Davies, N., Smith, M., Steggles, P.: Towards a Better Understanding of Context and Context-Awareness. In: Gellersen, H.-W. (ed.) HUC 1999. LNCS, vol. 1707, p. 304. Springer, Heidelberg (1999)
17. Yang, S.J.H.: Context Aware Ubiquitous Learning Environments for Peer-to-Peer Collaborative Learning. Educational Technology & Society 9(1), 188–201 (2006)
18. Hwang, G.-J., Yang, T.-C., Tsai, C.-C., Yang, S.J.H.: A context-aware ubiquitous learning environment for conducting complex science experiments. Computers & Education 53(2), 402–413 (2009)
19. Blöckner, M., Danti, S., Forrai, J., Broll, G., De Luca, A.: Please touch the exhibits!: using NFC-based interaction for exploring a museum. In: Proceedings of the 11th International Conference on Human-Computer Interaction with Mobile Devices and Services, MobileHCI 2009, vol. 71, pp. 1–2 (2009)
20. Tan, Q., Kinshuk, Kuo, Y.-H., Jeng, Y.-L., Wu, P.-H., Huang, Y.-M., Liu, T.-C., et al.: Location-Based Adaptive Mobile Learning Research Framework and Topics. In: International Conference on Computational Science and Engineering, CSE 2009, vol. 1, pp. 140–147 (2009)
21. Saccol, A.Z., Kich, M., Schlemmer, E., Reinhard, N., Barbosa, J.L., Hahn, R.: A Framework for the Design of Ubiquitous Learning Applications. In: 42nd Hawaii International Conference on System Sciences, HICSS 2009, pp. 1–10 (2009)
22. Driver, C., Clarke, S.: An application framework for mobile, context-aware trails. Pervasive and Mobile Computing 4(5), 719–736 (2008)
23. Padovitz, A., Loke, S., Zaslavsky, A.: The ECORA framework: A hybrid architecture for context-oriented pervasive computing. Pervasive and Mobile Computing 4(2), 182–215 (2008)
24. Nebusens, n-Core®: A Faster and Easier Way to Create Wireless Sensor Networks. n-Core® (2012), `http://www.n-core.info` (last access: January 18, 2012)

Application of Soft Computing in the Assessment of Comprehensive Skills of First Year Dental Student

Ignacio Aliaga[1], Vicente Vera[1], Héctor Casado[2], Cristina González Losada[1], Álvaro Enrique García Barbero[1], and Emilio S. Corchado Rodríguez[2]

[1] Odontology Faculty, University Complutense of Madrid, Madrid, Spain
ialia01@estumail.ucm.es, {vicentevera,aegarcia}@odon.ucm.es
[2] Departamento de Informática y Automática, University of Salamanca, Salamanca, Spain
{hectorscasa,escorchado}@usal.es

Abstract. Professional education in dentistry exists to educate good dentists who are equipped and committed to helping society gain the benefits of oral health. In achieving this objective, dental educators acknowledge that student dentists must acquire the complex knowledge base and sophisticated perceptual-motor skills of dentistry. This study examined the validity of both cognitive and noncognitive factors used for the evaluation of the students in the reputed dental school of the Complutense University of Madrid. Interest in personality measurement and the prediction offered by personality measures has escalated and may be applied during the evaluation of dental students. Therefore, the study also assessed whether the addition of a personality measure would increase the validity of predicting performance beyond that achieved by PAU (the university admissions test) and the practical training period. The real data collected from a set of first year dental students was analyzed by using statistical and soft computing models. Results suggest that the questionnaire may be useful in identifying specific behavioural characteristics deemed important for success in dental training.

Keywords: Unsupervised Neural and Exploratory Projection Techniques, Evaluation of dental students, Professional education in dentistry, Dental milling.

1 Introduction

An instructor's awareness of students, of the socioeducational context, and of the inherent dynamics within classroom groups is important in the definition of course content and in the development and design of the curriculum. The timely identification of structures, hierarchies and subgroups in a group of students means the instructor can focus on follow up work and make individual or group changes so as to optimize the learning/teaching process. It is not an easy task, especially with large groups and with study modules that have few teaching hours. As an objective contribution to that awareness, conventional assessment tools are available to

L. Uden et al. (Eds.): Workshop on LTEC 2012, AISC 173, pp. 23–33.
springerlink.com

the instructor, which are complemented by subjective observations based on professional experience and "wisdom" (classroom time, personal consultation, tutorials, etc.). Quality improvement systems are fundamentally based on objective measurements generated by conventional assessment models or generalizations drawn from student satisfaction surveys. All of these are conducive to positive outcomes in the teaching/learning process, but lack an immediacy that is desirable for decision-making in the classroom [1].

Practical assignments performed by students of Dentistry, such as the creation of methacrylate figures during the practical training period, could be considered a type of activity that can be objectively assessed by the application of measurement systems or metrics.

Analytical and multidimensional data visualization techniques are often applied in a range of professional contexts [2–6]. They provide tools that are intended to facilitate the interpretation of results, and thus improve the effectiveness of decision-making that might affect the progress of a business. It would appear reasonable for computing professionals involved in teaching tasks to take advantage of those same improvements [7].

Consequently, this study applied projection techniques to multivariate data to obtain a 2D representation, simplifying the dataset but looking for the "most interesting" directions, with regard to the directions that highlight specific aspects in the dataset. Principal Component Analysis (PCA) [8, 9] was used, as well as a neuronal model of Exploratory Projection Pursuit (EPP), Maximum-Likelihood Hebbian Learning (MLHL) [10, 11].

The objective of this study was to discover student groupings based on the grades received for the course "Introduction to Dentistry", which may not be easily determined by means of quotidian contact in the classroom or conventional assessment techniques. These observations may reveal individual or group non-desirable discordant practices, allowing instructors to focus on the individuals and determine different adaptive teaching strategies based on their own experience. The case study also checked if the observed groupings might have an academic origin, but produced negative results. A comparative study was done on the results obtained from classic code metrics and no valuable observation was obtained from those graphs.

2 Dimensionality Reduction Visualization for Data Analysis

Projection methods project high-dimensional data points onto lower dimensions in order to identify "interesting" directions in terms of any specific index or projection. Such indexes or projections are, for example, based on the identification of directions that account for the largest variance of a dataset (such as Principal Component Analysis (PCA) [8]) or the identification of higher order statistics such as the skew or kurtosis index, as in the case of Exploratory Projection Pursuit (EPP) [9]. Having identified the interesting projections, the data is then projected onto a lower dimensional subspace plotted in two or three dimensions, which makes it possible to examine its structure with the naked eye. The remaining dimensions are discarded as they mainly relate to a very small percentage of the information or the dataset structure. In that way, the structure identified through a multivariable dataset may be visually analysed with greater ease.

A combination of these types of techniques together with the use of scatter plot matrices constitute a very useful visualization tool to investigate the intrinsic structure of multidimensional datasets, allowing experts to study the relationship between different components, factors or projections, depending on the technique that is used.

2.1 The Unsupervised Connectionist Model

The standard statistical EPP method [9] provides a linear projection of a dataset, but it projects the data onto a set of basic vectors which best reveal the interesting structure in data; interestingness is usually defined in terms of how far the distribution is from the Gaussian distribution.

One neural implementation of EPP is Maximum-Likelihood Hebbian Learning (MLHL) [11], which identifies interestingness by maximising the probability of the residuals under specific probability density functions that are non-Gaussian.

Considering an N-dimensional input vector (*x*), and an M-dimensional output vector (*y*), with W_{ij} being the weight (linking input *j* to output *i*), then MLHL can be expressed as:

1. Feed-forward step:

$$y_i = \sum_{j=1}^{N} W_{ij} x_j, \forall i \tag{1}$$

2. Feedback step:

$$e_j = x_j - \sum_{i=1}^{M} W_{ij} y_i, \forall j \tag{2}$$

3. Weight change:

$$\Delta W_{ij} = \eta . y_i . sign(e_j) |e_j|^{p-1} \tag{3}$$

Where: η is the learning rate and *p* a parameter related to the energy function [11].

3 Measurement and Comparison of Dentistry Skills

The aim of this study is to classify the skills of first year students of dentistry when creating methacrylate figures during this **Dental Aptitude Test.** The study seeks to facilitate the identification of divergent or non-desirable situations in the educational process. Interest in measurements and the prediction offered by these measures has escalated and may be applied during the evaluation of dental students [12].

The students carve ten methacrylate figures, using rotatory systems, applying two different speeds (V1 and V2). A total of twenty figures.V1 (low speed) rotates at a speed of 10-60,000 revolutions per minute (rpm), while V2 (turbine or high speed) rotates at a speed of 250,000 rpm.

Seven of the figures made by students are easily created, while the remainder, which have several planes, involve a bit more difficultly.

3.1 Description of the Test on Students

These tests provide a comprehensive analysis of a student's ability with regard to their peer group and graduation year. The development of a student's learning during the course of the academic year, compared to other academic years or to other students in the same year, can be measured with this test.

The student works on a methacrylate sheet. Students will differentiate two of the most important materials in dentistry, low speed and high speed. They will use the low speed (10.000-60.000 RPM) to carve a completed set of ten figures drawn by each student in the methacrylate sheet (see Fig 1). After completing the first part of the practical training period, students will start carving the second part, which is basically a second round of the same figures, but this time using the high speed (150.000-250.000 RPM). Using this second instrument involves more difficulty as the bur is spinning faster than the low speed, and requires more manual skills.

Both parts will be completed in 90 minutes. Upon completion of training the students must submit their methacrylate block with all the figures carved on it.

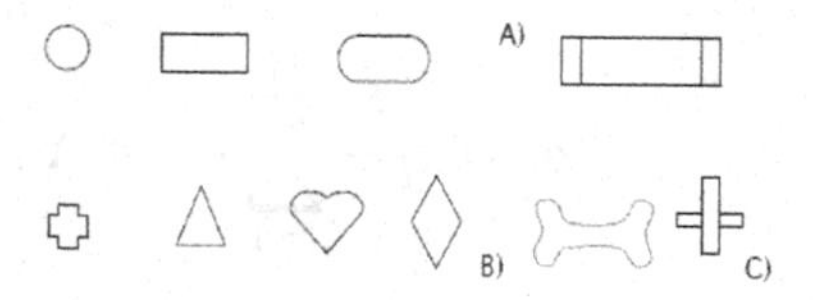

Fig. 1 Figures carved by the students

The steps needed for students to complete their training are as follows:

- Assemble the low speed and turbine properly and test its operation. Recognize and properly mount the milling cutters.
- The student will create figures in a methacrylate block (see Fig 2). Previously, the students have to draw the outline of what to carve with a pen and a ruler.
- Figures will have an approximate size of 1 cm (on the longest side) and a depth of about 3-4 mm, except figures A, B and C (see Fig 1).
- The cavity design must be clear and free of defects, and the walls must be smooth and perpendicular to the surface of the methacrylate sheet. The floor must be parallel to the surface and as smooth as possible.

- Cavities are initially carved with a low speed, and after with a turbine, depending on the skills acquired by students.
- Figures A, B and C (see Fig 1) will be the last ones created because their difficulty is greater, as they have 2 planes.
- Students can repeat figures as often as they like, so they can use all but one side of the methacrylate block.
- If the milling cutter is dulled with resin, it can be cleaned by spinning briefly and gently against a small block of wood or another milling cutter cleaner.
- At the end of the training period the students must submit their methacrylate block with all the figures carved on it (see Fig 2).

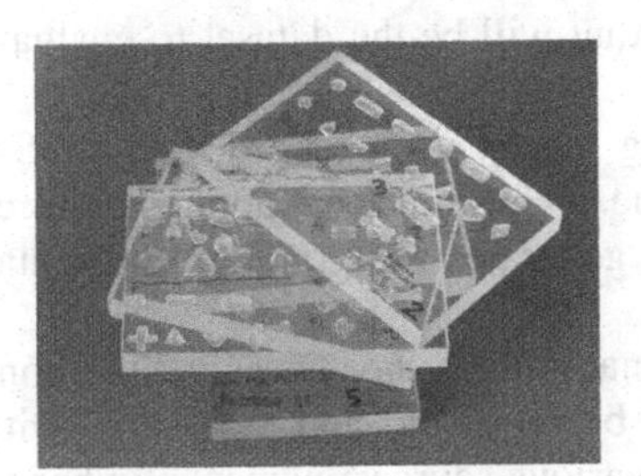

Fig. 2 Real methacrylate sheet with figures carved on it

3.2 A Real Case Study

The real case study is based on a survey of 79 first year dental students (24 students were eliminated from the initial 103 students, because they did not participate in the practical training). The information analyzed for each student was based on the following 88 variables. The first eight variables are:

- Age of the student (integer value).
- Sex of the student (integer value).
- Mark obtained by the student in the university entrance exam (PAU). (Decimal value between 0 and 14).
- Previous experience gained by the student. The students may have had professional experience as nurse, dental technician, hygienist, dental technician and hygienist, or lack of previous work experience.
- Mark obtained by the student in the theoretical exam of the subject (Decimal value between 0 and 10).
- Mark obtained by the students in the practical part of the subject (Decimal value between 0 and 10).
- Group class of the student (integer value between 1 and 4).
- Number of figures carved by the student (integer value between 0 and 20).

The following eighty variables (twenty figures with four variables each) are the evaluations of the different parts of the figures (graded between 0 and 5). The way to interpret these variables is as follows: 'x' indicates the figure number and can range from 1 to 10, 'y' indicates the speed used to carve the figure (Low Speed (1), High speed (2)), and 'z' indicates the evaluator who examines the test (1 or 2):

- **Fx_Vy_Ez_WALL:** evaluate the quality of the walls of the figure created by the student.
- **Fx_Vy_Ez_DEPTH:** evaluate the quality of the depth of the figure created by student.
- **Fx_Vy_Ez_EDGES:** evaluate the quality of the edges of the figure created by student.
- **Fx_Vy_Ez_FLOOR:** evaluate the plain and irregularities presented on the floor of the figure created by the student.

All these real data are collected in a document that will be the dataset to evaluate and study.

These datasets, along with the corresponding labels, were recorded in a CSV format text file that was used as input data in the programme that applies the previously described reduction treatment and that generates the graphic representations.

Data preparation, performed on a conventional spread sheet, was a time consuming task, as a great amount of data had to be reordered and associated with academic management information taken from various sources: names, number of students completing the practical assignments, marks, etc.

4 Data Analysis

PCA identified two clearly separate clusters; G_1 and G_2 (see Fig 3).

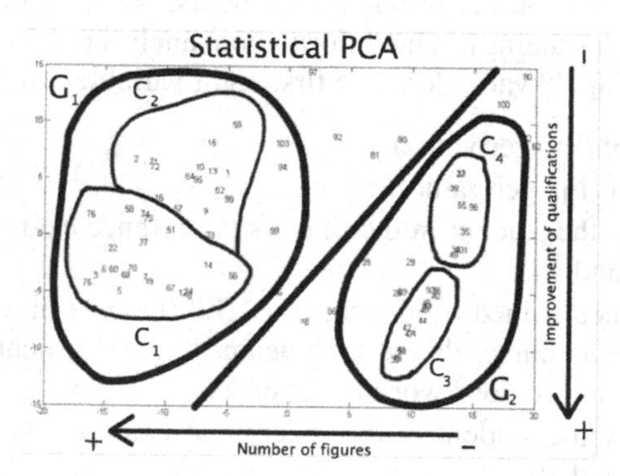

Fig. 3 Statistical PCA

- **G_1:** this cluster represents students with high marks in the university entrance exam, good or very good marks on the theoretical and practical issues of the subject area, and a large number of figures carved. This cluster is composed by two sub-clusters: C_1 and C_2.

- **G_2:** this cluster represents students with decent marks in theoretical and practical issues of the subject area, and a small number of figures carved. This cluster is composed of two sub-clusters: C_3 and C_4.

These clusters are divided into the following sub-clusters:

- C_1 represents young students (see Table 1) characterized with high marks in the university entrance exam, without previous professional experience, good marks in both theory and practice.
- C_2 represents young students (see Table 1) with no previous professional experience, with good marks in the theoretical part of the subject area and average marks in the practical part. These students have carved many figures, but less than students belonging to the previous cluster (C_1).
- C_3 represents students (see Table 1) with decent marks, both in the theoretical and practical parts of the subject area, and have succeeded in carving an average number of figures (about half), mostly with low speed. Most of these students are in group number 2.
- C_4 represents students (see Table 1) that have been able to successfully carve 7 figures. Their marks in theory and practice are varied.

Table 1 Cluster Classification (PCA)

Clusters	Students belonging to each cluster
C1	3, 5, 6, 7, 8, 12, 19, 22, 24, 51, 58, 60, 67, 68, 70, 73, 74, 75, 76, 77
C2	1, 2, 9, 10, 13, 15, 16, 23, 52, 57, 59, 71, 72, 84, 85, 98
C3	4, 32, 36, 40, 41, 42, 43, 44, 45, 46, 47, 49, 50, 93
C4	33, 34, 35, 37, 39, 48, 55, 96, 101

4.1 Maximum Likelihood Hebbian Learning (MLHL)

Having analyzed the data using PCA, MLHL is then applied to find possible improvements in the classification of the samples (students).

Once the test using MLHL is performed (see Fig 4), the best result is obtained by applying the following parameters: 3 neurons in the output layer (m), 100000 iterations, 0.006 as learning rate, and 1.1 as p.

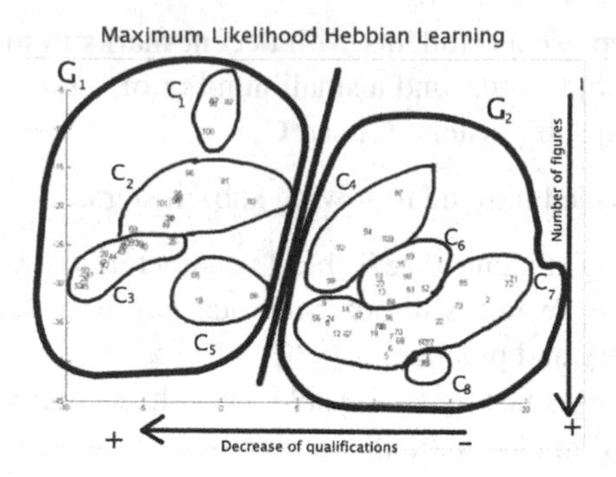

Fig. 4 MLHL Analysis with m=3, iters=100000, lrate=0.006 and p=1.1

There are two clusters, as in PCA, G_1 and G_2. In this case the first cluster (G_1) is composed of four sub-clusters C_1, C_2, C_3, and C_5, and is characterized by an average mark of less than 1.8 for the figures performed at high speed (V2). The second cluster (G_2) is composed of four sub-clusters (C_4, C_6, C_7 and C_8) and is characterized by an average mark greater than 1.8 for the figures carved at high speed (V2).

In order to facilitate the reading of the document, the following equivalences will be adopted from this point forward:

- Overall average: average of marks for figures created by a student.
- Average V1: average of marks for figures carved by a student using low speed.
- Average V2: average of marks for the figures created by a student using turbine (high speed).

The clusters that describe the data classification are:

- C_1 represents students (see Table 2) whose overall average is between 0.0 and 0.5. This cluster is also characterized by an average V1 between 0 and 0.95 and an average V2 of mostly 0.
- C_2 represents students (see Table 2) who have been able to successfully perform seven figures and whose overall average falls between 0.8 and 1.5. This cluster is also characterized by an average V1 between 1.7 and 2.95 and an average V2 of mostly 0.
- C_3 represents students (see Table 2) who have an overall average between 1.5 and 2.1. Students in this cluster, are characterized by having made an average number of figures (between 8 and 13), most of which are carved with low speed (V1). This cluster is also characterized by an average V1 between 3 and 4.15 and an average V2 of mostly 0.
- C_4 represents students (see Table 2) who are similar to those belonging to the previous cluster, as the overall average of students falls within the same range, and the average V1 is a subinterval of students in cluster C_2, namely between 1.64 and 2.48. However, they differ in the average V2, because it falls within a completely different range (between 1.88 and 2).

- **C_5 represents students (see Table 2) with an overall average similar to those in clusters C_3 y C_4 and who have created 13 or 14 figures. The overall V1 is a subinterval of the overall V1 of C_3 students, but the overall V2 is between 0.9 and 1.7.**
- **C_6 represents students (see Table 2) with an overall average between 2.1 and 2.8 and who have created 14-20 figures. Although this range is included in the interval of the previous cluster, they are not the same because the overall V1 is between 2.4 and 3.5 and the overall V2 is between 2.4 and 3.15.**
- **C_7 represents students (see Table 2) with an overall average between 2.8 and 3.75. The average V1 is between 3.1 and 3.7 and the average V2 is between 2.05 and 3.9. This is the largest cluster detected.**
- **C_8 represents students (see Table 2) with the best overall average (between 3.9 and 4.1). Likewise, the average for both V1 and V2 is greater than 3.9. These students have been able to perform (almost) all the figures.**

The following table shows which students belong to each cluster:

Table 2 Cluster Classification (MLHL)

Clusters	Students belonging to each cluster
C1	62, 82, 90, 100
C2	33, 34, 35, 37, 39, 48, 55, 69, 80, 81, 96, 101
C3	4, 25, 26, 29, 32, 36, 40, 41, 42, 43, 44, 45, 46, 47, 49, 50, 93
C4	92, 94, 97, 99, 103
C5	18, 65, 86
C6	1,10,13,15,23,51,52,59,84, 98
C7	2, 5, 6, 7, 8, 9, 12, 14, 16, 19, 22, 24, 56, 57, 58, 60, 67, 68, 70, 71, 72, 73, 74, 77, 85
C8	3, 75, 76

5 PCA and MLHL Comparison

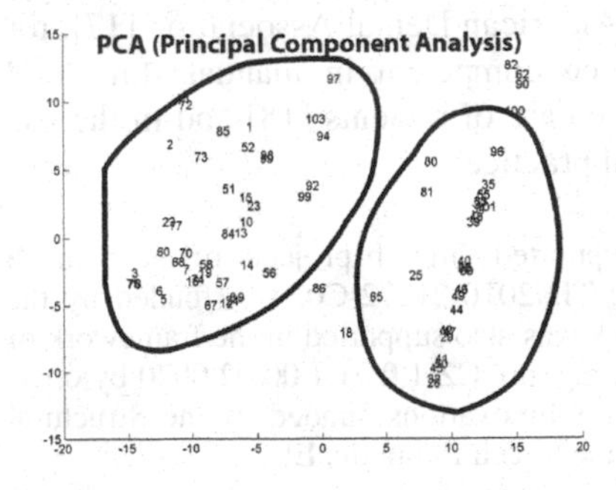

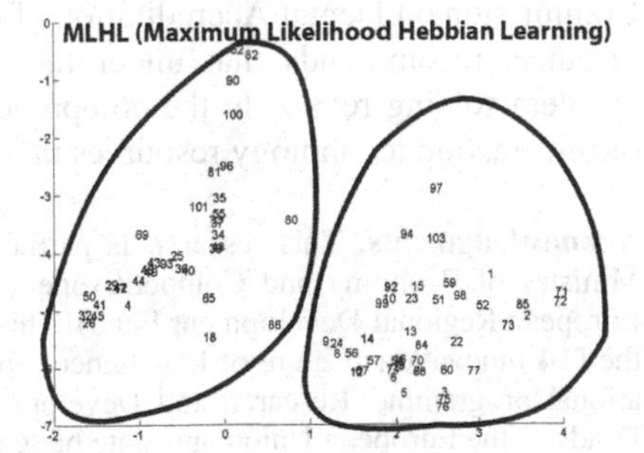

Fig. 5 PCA and MLHL Comparison

As seen in Figure 5, and shown in previous sections, the study based on the use of PCA provides a valid initial classification of the real dataset, but it can certainly be improved by using MLHL.

Using MLHL makes it possible to obtain a better visual classification of the internal structure of the dataset as it manages to group the samples that have more in common and thus provide a more refined classification.

As seen in previous sections, PCA analysis classifies students into two clusters, each of which contains two sub-clusters On the other hand, while MLHL analysis also classifies the students into two clusters each of these clusters contains four sub-clusters, which greatly improves the ability to refine the classification of students. The improvements derived from the use of MLHL studies are due to the use of higher order statistics, such as skew or kurtosis index, which are not used by PCA.

6 Conclusions

The most significant finding from this study is that there are different sets of students based on clinical and academic components of dental training. The evaluation of these variables is based on performance measurements related to manual skills and the students' academic record. These evaluations will predict the academic success of most students at dental school [13]. Dental students need to have both academic and clinical skills [14].

Certain variables may predict progress in clinical components of the program, while others may help to distinguish those students who will pass from those likely to fail.

This study also supports the use of personality measures in the evaluation process of the student and the importance of assessing their behavioural characteristics, although more work is needed in this regard. The relationship between the PAU, the practical training experience, and the number of figures created, raises the possibility that an appropriate set of personality measures would help to predict clinical success. There is no significant relationship with prior professional experience.

From an educational perspective, educators and schools have proposed changes in their objectives and competencies to enhance the skills of undergraduates in evidence-based practice and the use of information technology [15, 16]. The Commission on Dental Accreditation of the American Dental Association [17], for instance, recommends that undergraduates be competent in manual skills and problem solving related to the comprehensive care of patients [18] and in the use of information technology resources in dental practice.

Acknowledgments. This research is partially supported through projects of the Spanish Ministry of Economy and Competitiveness [ref: TIN2010-21272-C02-01 (funded by the European Regional Development Fund). This work was also supported in the framework of the IT4 Innovations Centre of Excellence project, reg. no. CZ.1.05/1.1.00/02.0070 by operational programme 'Research and Development for Innovations' funded by the Structural Funds of the European Union and state budget of the Czech Republic, EU.

References

1. Behar-Horenstein, L.S., Schneider-Mitchell, G., Graff, R.: Faculty Perceptions of a Professional Development Seminar. J. Dent. Educ. 72, 472–483 (2008)
2. Fiori, S.: Visualization of Riemannian-manifold-valued elements by multidimensional scaling. Neurocomputing 74, 983–992 (2011)
3. González-Navarro, F.F., Belanche-Muñoz, L.A., Romero, E., Vellido, A., Julià-Sapé, M., Arús, C.: Feature and model selection with discriminatory visualization for diagnostic classification of brain tumors. Neurocomputing 73, 622–632 (2010)
4. Corchado, E., Herrero, Á.: Neural visualization of network traffic data for intrusion detection. Applied Soft Computing 11, 2042–2056 (2011)
5. Corchado, E., Yin, H. (eds.): IDEAL 2009. LNCS, vol. 5788. Springer, Heidelberg (2009)
6. Corchado, E., Perez, J.: A three-step unsupervised neural model for visualizing high complex dimensional spectroscopic data sets. Pattern Analysis & Applications 14, 207–218 (2011)
7. Dorothy McComb, B.D.S.: Class I and Class II silver amalgam and resin composite posterior restorations: teaching approaches in Canadian faculties of dentistry. J. Can. Dent. Assoc. 71, 405–406 (2005)
8. Hotelling: Analysis of a complex of statistical variables into principal components. Journal of Educational Psychology 24, 417–441 (1933)
9. Friedman, J.H., Tukey, J.W.: A Projection Pursuit Algorithm for Exploratory Data Analysis. IEEE Transactions on Computers C-23, 881–890 (1974)
10. Corchado, E., Pellicer, M.A., Borrajo, M.L.: A maximum likelihood Hebbian learning-based method to an agent-based architecture. International Journal of Computer Mathematics 86, 1760–1768 (2009)
11. Corchado, E., MacDonald, D., Fyfe, C.: Maximum and Minimum Likelihood Hebbian Learning for Exploratory Projection Pursuit. Data Mining and Knowledge Discovery 8, 203–225 (2004)
12. Behar-Horenstein, L.S., Mitchell, G.S., Dolan, T.A.: A Case Study Examining Classroom Instructional Practices at a U.S. Dental School. J. Dent. Educ. 69, 639–648 (2005)
13. Mofidi, M., Strauss, R., Pitner, L.L., Sandler, E.S.: Dental Students' Reflections on Their Community-Based Experiences: The Use of Critical Incidents. J. Dent. Educ. 67, 515–523 (2003)
14. Fukushima, M., Iwaku, M., Setcos, J.C., Wilson, N.H., Mjör, I.A.: Teaching of posterior composite restorations in Japanese dental schools. International Dental Journal 50, 407–411 (2000)
15. Coy, K., McDougall, H., Sneed, M.: Issues Regarding Practical Validity and Gender Bias of the Perceptual Abilities Test (PAT). J. Dent. Educ. 67, 31–37 (2003)
16. Sandow, P.L., Jones, A.C., Peek, C.W., Courts, F.J., Watson, R.E.: Correlation of Admission Criteria with Dental School Performance and Attrition. J. Dent. Educ. 66, 385–392 (2002)
17. Lynch, C.D., McConnell, R.J., Wilson, N.H.F.: Teaching of posterior composite resin restorations in undergraduate dental schools in Ireland and the United Kingdom. European Journal of Dental Education 10, 38–43 (2006)
18. Liew, Z., Nguyen, E., Stella, R., Thong, I., Yip, N., Zhang, F., Burrow, M.F., Tyas, M.J.: Survey on the teaching and use in dental schools of resin-based materials for restoring posterior teeth. Int. Dent. J. 61, 12–18 (2011)

An Overview of E-Learning in Cloud Computing

A. Fernández[1], D. Peralta[2], F. Herrera[2], and J.M. Benítez[2]

[1] Dept. of Computer Science. University of Jaén,
Jaén, Spain
alberto.fernandez@ujaen.es

[2] Dept. of Computer Science and Artificial Intelligence, CITIC-UGR (Research Center on Information and Communications Technology). University of Granada,
18071 Granada, Spain
{dperalta,herrera,jmbs}@decsai.ugr.es

Abstract. E-Learning is the topic related to the virtualized distance learning by means of electronic communication mechanisms, specifically the Internet. They are based in the use of approaches with diverse functionality (e-mail, Web pages, forums, learning platforms, and so on) as a support of the process of teaching-learning. The Cloud Computing environment rises as a natural platform to provide support to e-Learning systems and also for the implementation of data mining techniques that allow to explore the enormous data bases generated from the former process to extract the inherent knowledge, since it can be dynamically adapted by providing a scalable system for changing necessities along time.

In this contribution, we give an overview of the current state of the structure of Cloud Computing for applications on e-learning. We provide details of the most common infrastructures that have been developed for such a system, and finally we present some examples of e-learning approaches for Cloud Computing that can be found in the specialized literature.

1 Introduction

The Electronic Learning, better known as E-Learning [13], is defined as an Internet-enabled learning. Components of e-Learning can include content of multiple formats, management of the learning experience, and an online community of learners, content developers and experts. The study summarized the main advantages, which include flexibility, convenience, easy accessibility, consistency and its repeatability.

With Information Technologies (IT), there is a growing trend regarding the research and exploitation of this kind of e-Learning platforms. There exist several initiatives at different educative levels, from which some examples are the Khan Academy[1], the Virtual Learning Center of Granada University (CEVUG-UGR),

[1] www.khanacademy.org

L. Uden et al. (Eds.): Workshop on LTEC 2012, AISC 173, pp. 35–46.
springerlink.com

the Open University of Catalonia, the MIT Open CourseWare, or the "Free Online Course" of the Standford University.

The virtual courses that are supported by the e-Learning approach favors the achievement of a higher impact for the educative framework than those of the classical attendance group. As an example, in the first edition of the "Machine Learning" course of Stanford[2] more than 160,000 worldwide students were registered. These dimensions affects different issues; on the one hand, the infrastructure provisions that are necessary to give a concurrent service for that amount of students clearly exceed the capabilities of a conventional web server. Furthermore, the demand of the teaching resources usually vary in a dynamic and very quick way, and presents high peaks of activity. To attend requests during these periods of time without other system services to be resented, it will be necessary to prepare a quite superior infrastructure than that required for the regular working of the learning institution. An alternative would be to provide those services depending on the demand and only paying for the resources that are actually used. The answer to these necessities is the Cloud Computing environment.

Cloud Computing [3, 18] is a computation paradigm in which the resources of an IT system are offered as services, available to the users through net connections, frequently the Internet. It is a model of provision of IT services offered through a catalog that answers to the necessities of the user in a flexible and adaptive way, only billing for the actual usage that is made. Therefore, two of the distinctive features of this paradigm are, on the one hand, the use of resources under demand and, on the other hand, the transparent scalability in such a way that the computational resources are assigned in a dynamical and accurate manner when they are strictly necessary, without the requirement of a detailed understanding of the infrastructure from the user's point of view.

With these characteristics, the Cloud platforms arise as accurate alternatives to traditional computer centers. They represent a significative alternative versus the acquisition and maintenance of the computer centers.

Additionally, the e-learning platforms of the large dimensions which we mentioned above generate extensive registers of interaction among students-platform-teachers. These data bases contain significative information not defined in a precise way. Data Mining techniques must be applied to extract this information [23, 17]. Therefore "Educational Data Mining"[3] comes up, being this a discipline whose object of interest is the development of new methodologies to explore the data that are generated in the activity of the educational systems (mainly those with a technological base) and the application of such methods to achieve a better understanding of the behaviour of the students, and how to design procedures and material that ease the learning process.

[2] `http://www.ml-class.org/course/class/index`
[3] `http://www.educationaldatamining.org/`

In clear connection with this process we may find the Intelligent Tutoring Systems[4] which are computer based systems to support the teaching-learning process. Usually, they are intelligent systems able to drive the learning process of the student providing him/her feedback based on the progress of the student and the results of periodical tests. The process of "Educational Data Mining" interacts with an Intelligent Tutoring System by extending and refining its knowledge base. Taking into account the dimensions and growing capacity of the computational resources (stable storage, memory and CPUs) a Cloud platform is also a natural structure for the implementation of data mining techniques and their application to growing data-sets (Big Data). However, many of the data mining techniques do not have an adequate scalability. This is an aspect that grows in importance and that have attracted the interest of researchers and companies.

In order to overview all these aspects, this contribution is arranged as follows. In Section 2 we introduce the main concepts on Cloud Computing, including its infrastructure and main layers. Next, Section 3 presents the features of the e-Learning approach, stressing the advantages of the migration of such a system to a Cloud Computing environment and showing some examples of real applications of this kind. Finally, the main concluding remarks are given in Section 4.

2 Basic Concepts on Cloud Computing

We may define an SOA [15] as an integration platform based on the combination of a logical and technological architecture oriented to support and integrate all kind of services. In general, a "Service" in the framework of Cloud Computing is a task that has been encapsulated in a way that it can be automated and supplied to the clients in a consistent and constant way. Any component can be considered as a service, from entities closest to hardware such as the storage space or the computational time, to software components aimed at authenticating a user or to manage the mailing, the management of a data base or the monitoring of the use of the system resources.

In this section we will give a brief introduction to the Cloud Computing environment, first describing its main features, next by presenting the layers in which this platform is built of, and finally pointing out several technological difficulties that should still be addressed to improve the quality of this paradigm.

2.1 Introduction to Cloud Computing

The philosophy of Cloud Computing mainly implies a change in the way of solving the problems by using computers. The design of the applications is based upon the use and combination of services. On the contrary that occurs in more traditional approaches, i.e. grid computing, the provision of the functionality relays on this use

[4] `http://aaai.org/AITopics/IntelligentTutoringSystems`

and combination of services rather than the concept of process or algorithm. The idea behind this is that grid computing mainly focuses on high performance computing whereas Cloud Computing offers both standard and intensive computation. Additionally, Cloud offers more services than grid computing, i.e web hosting, multiple Operating systems, DB support and much more. Finally, grids tends to be more loosely coupled, heterogeneous, and geographically dispersed compared to conventional cluster computing systems.

Clearly, this brings advantages in different aspects, for example the scalability, reliability, and so on, where an application, in the presence of a peak of resources' demand, because of an increase of users or an increase of the data that those provide, can still give an answer in real time since it can get more instances of a determinate service; the same occurs in the case of a fall of the demand, for which it can liberate resources, all of these actions in a transparent way to the user.

The main features of this architecture are its loose coupling, high inter-operativity and to have some interfaces that isolate the service from the implementation and the platform. In an SOA, the services tend to be organized in a general way in layers or levels (not necessarily with strict divisions) where normally, some modules use the services that are provided by the lower levels to offer other services to the superior levels. Furthermore, those levels may have different organization structure, a different architecture, etc.

2.2 *Cloud Computer Layers*

There exists different categories in which the service oriented systems can be clustered. One of the most used criteria to group these systems is the abstraction level that offers to the system user. In this manner, three different levels are often distinguished , as we can observe in Figure 1. In the remainder of this section, we will first describe each one of these three levels, providing the features that defines each one of them and some examples of the most known systems of each type. Next we will present some technological challenges that must be taken into account for the development of a Cloud Computing system.

- Infrastructure as a Service (IaaS): IaaS is the supply of hardware as a service, that is, servers, net technology, storage or computation, as well as basic characteristics such as Operating Systems and virtualization of hardware resources [8]. Making an analogy with a monocomputer system, the IaaS will correspond to the hardware of such a computer together with the Operating System that take care of the management of the hardware resources and ease the access to them.
- Platform as a Service (PaaS): At the PaaS level, the provider supplies more than just infrastructure, i.e. an integrated set of software with all the stuff that a developer needs to build applications, both for the developing and for the execution stages. In this manner, a PaaS provider does not provide the infrastructure directly, but making use of the services of an IaaS it presents the tools that a developer needs, having an indirect access to the IaaS services and, consequently, to the infrastructure [8].

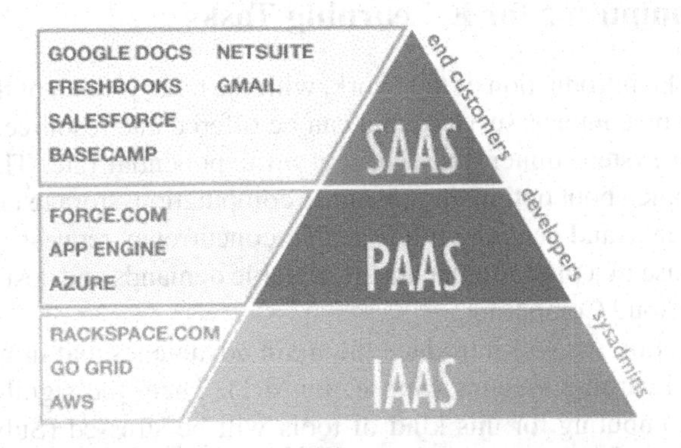

Fig. 1 Illustration of the layers for the Services Oriented Architecture

- Software as a Service (SaaS): In the last level we may find the SaaS, i.e. to offer software as a service. This was one of the first implementations of Cloud services. It has its origins in the host operations carried out by the Application Service Providers, from which some companies offered to others the applications known as Customer Relationship Managements [5].

2.3 Technological Challenges in Cloud Computing

Cloud computing has shown to be a very effective paradigm according to its features such as on-demand self-service since the customers are able to provision computing capabilities without requiring any human interaction; broad network access from heterogeneous client platforms; resource pooling to serve multiple consumers; rapid elasticity as the capabilities appear to be unlimited from the consumer's point of view; and a measured service allowing a pay-per-use business model. However, there are also some weak points that should be taken into account. Next, we present some of these issues:

- Security, privacy and confidence: Since the data can be distributed on different servers, and "out of the control" of the customer, there is a necessity of managing hardware for computation with encoding data by using robust and efficient methods. Also, in order to increase the confidence of the user, several audits and certifications of the security must be performed.
- Availability, fault tolerance and recovery: to guarantee a permanent service (24x7) with the use of redundant systems and to avoid net traffic overflow.
- Scalability: In order to adapt the necessary resources under changing demands of the user by providing an intelligent resource management, an effective monitorization can be used by identifying a priori the usage patterns and to predict the load in order to optimize the scheduling.
- Energy efficiency: It is also important to reduce the electric charge by using microprocessors with a lower energy consumption and adaptable to their use.

3 Cloud Computing for E-Learning Tasks

As we stated in the introduction of this work, with the huge growth of the number of students, education contents, services that can be offered and resources made available, e-Learning system dimensions grow at an exponential rate. The challenges regarding this topic about optimizing resource computation, storage and communication requirements, and dealing with dynamic concurrency requests highlight the necessity of the use of a platform that meets scalable demands and cost control. This environment is Cloud Computing.

Along this section we will introduce the main advantages and drawbacks to be addressed for e-Learning systems (Subsection 3.1). Then, the significance of selecting Cloud Computing for this kind of tools will be stressed (Subsection 3.2). The organization and infrastructure necessary for the virtual platform is described next (Subsection 3.3). Finally, we will review some of the e-Learning applications that have been already developed within the Cloud Computing platform (Subsection 3.4).

3.1 Current Challenges of E-Learning Systems

Among the learning technologies, web-based learning offers several benefits over conventional classroom-based learning. Its biggest advantages are the reduced costs since a physical environment is no longer required and therefore it can be used at any time and place for the convenience of the student. Additionally, the learning material is easy to keep updated and the teacher may also incorporate multimedia content to provide a friendly framework and to ease the understanding of the concepts. Finally, it can be viewed as a learner-centered approach which can address the differences among teachers, so that all of them may check the confidence of their material to evaluate and re-utilize common areas of knowledge [9].

However, there are some disadvantages that must be addressed prior to the full integration of e-Learning into the academic framework. Currently, e-Learning systems are still weak on scalability at the infrastructure level. Several resources can be deployed and assigned just for specific tasks so that when receiving high workloads, the system need to add and configure new resources of the same type, making the cost and resource management very expensive.

This key issue is also related to the efficient utilization of these resources. For example, in a typical university scenario, PC labs and servers are under-utilized during the night and semester breaks. In addition, these resources are on high demands mainly towards the end of a semester, following a dynamic rule of use. The physical machines are hold even when they are idle, wasting its full potential.

Finally, we must understand that there is a cost related to the computer (and building) maintenance, but that the educational center must pay for the site licensing, installation and technical support for the individual software packages [10].

3.2 *On the Suitability of Cloud Computing for E-Learning*

E-Learning in the Cloud can be viewed as Education Software-as-a-Service. Its deployment can be performed very quickly since the hardware requirements of the user are very low. Furthermore, as we stated previously, it lessens the burden of maintenance and support from the educational institution to the vendor, allowing them to focus on their core business, also obtaining the latest updates of the system without charges and sharing key resources using Web 2.0 technology.

In what follows, we summarize the consequences and implications regarding the development of e-Learning services within the Cloud Computing environment, as pointed out by Masud and Huang in [12]:

- **Accessed via Web:** It implies an ease of access since anywhere, any time and any one can access the application, greater demand for Web Development skills.
- **No client-side software needed:** Therefore, it has reduced costs for subscriber, as no installation, software maintenance, deployment and server administration costs, and a lower total cost of ownership, reduced time-to-value, fewer IT staff is needed by the institution.
- **Pay by subscription based on usage:** Which is suitable for Software Model Education market, and can gain access to more sophisticated applications.
- **SaaS server may support many educational institutions:** Since the application is running on a server farm, the scalability in inherent to the system. As student usage grows, the software performance will not degrade.
- **All subscriber data held on SaaS server:** Very high level of security is needed by SaaS provider in order to gain trust of subscribers and sophisticated multi-tenanted software architecture. The subscriber data is distributed between many providers and it must be integrated in order to gain overview of business, higher demand for system and data integrators.

Finally, several potential values of Cloud Computing for education as stressed by Ouf et al. in [14] include the following:

- No need for backing up everything to a thumb drive and transferring it from one device to another. It also means students can create a repository of information that stays with them and keeps growing as long as he wants them.
- Crash recovery is nearly unneeded. If the client computer crashes, there are almost no data lost because everything is stored in the cloud [16].
- Allow students to work from multiple Places (home, work, library ... etc), find their files and edit them through the cloud and browser-based applications can also be accessed through various devices (mobile, laptop and desk top computers, provided internet access is available) [2].
- Flexibility: Scale infrastructure to maximize investments. Cloud computing allows user to dynamically scale as demands fluctuate [6].
- Improved improbability : it is almost impossible for any interested person (thief) to determine where is located the machine that stores some wanted data (tests, exam questions, results) or to find out which is the physical component he needs to steal in order to get a digital asset [16].

- Virtualization: makes possible the rapid replacement of a compromised cloud located server without major costs or damages. It is very easy to create a clone of a virtual machine so the cloud downtime is expected to be reduced substantially.
- Centralized data storage: losing a cloud client is no longer a major incident while the main part of the applications and data is stored into the cloud so a new client can be connected very fast. Imagine what is happening today if a laptop that stores the examination questions is stolen.
- Monitoring of data access becomes easier in view of the fact that only one place should be supervised, not thousands of computers scattered over an extensive geographical area, for example. Also, the security changes can be easily tested and implemented since the cloud represents a unique entry point for all the clients [22].

3.3 Organization of the Cloud Computing Environment

The architecture of a Cloud Computing platform as depicted in Figure 2 is usually common to most e-Learning approaches on the Cloud. In the first layer we can observe the interface with the Cloud environment, which consists in several management subsystems for determining the current necessities of the user in terms of computational resources, being these the planner for the storage services, the management for distribution of the execution load among the virtual machines, a system administrator to monitor and to initiate activities of each layer, and a security component to ensure the privacy, recovery, integrity and security of user data and transactions, among others. The second layer represents the virtual machines implemented within the system. Finally, the third layer includes all the physical architecture of the system.

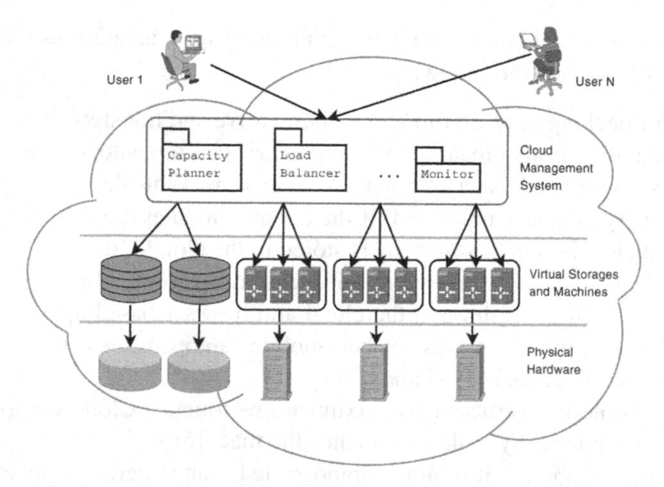

Fig. 2 Overview of a cloud architecture for e-Learning

Additionally, Liang and Yang describe in [11] the functions used in the Cloud IaaS and Saas which must be expected for developing such a system. These features can be observe in Figure 2 and are enumerated below:

- From the IaaS perspective:
  1. Storage management for the learning system and the users.
  2. Load Balance for all learning systems.
  3. Scaling management for virtual machines.
  4. Backup and Restore for the learning applications.
- From the SaaS perspective:
  1. Application Registry management for the commercial provides to register their applications.
  2. Application Server for managing and deploying the subscribed learning contents to the users.
  3. Account manage system for the authorized users.
  4. Virtual Desktop Deployment for providing the personalized desktop including the subscribed learning contents.
  5. Session Management for ensuring the Virtual Desktop used by the authorized user.
  6. Personalized management for managing the subscription of the favorite learning contents.

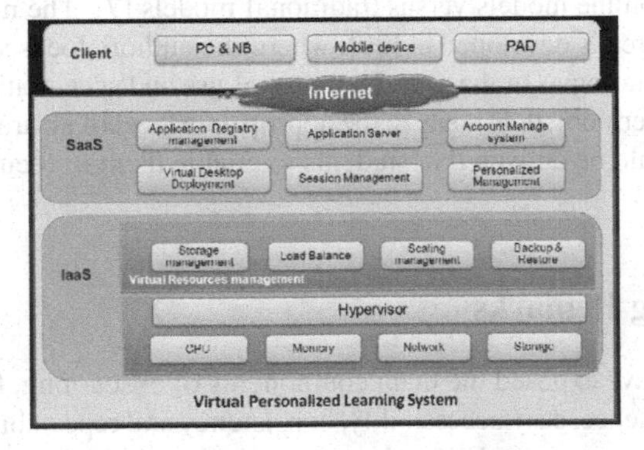

Fig. 3 The Architecture of the Virtual Personalized Learning Environment [11]

3.4 Applications of Cloud Computing for E-Learning

We must emphasize the necessity on setting the basis for a educational information infrastructure to alleviate the issues enumerated on the previous section. As we pointed out along this contribution, Cloud Computing may promote a new era of learning taking the advantage of hosting the e-Learning applications on a cloud and following its virtualization features of the hardware, it reduces the construction and maintenance cost of the learning resources.

At the present, the combination of cloud technologies and e-learning has been scarcely explored. Some relevant efforts to use IaaS cloud technologies in education focuses on the reservation of Virtual Machines to students for an specific time frame [21].

Another example of application that can be found in the specialized literature is BlueSky [4], whose architecture has several components aimed at the efficient provision and management of the e-Learning services, being able to pre-schedule resources for the hot contents and applications before they are actually needed, to safeguard the performance in concurrent access, although no details have been found with regard to how this is achieved. On the other hand, CloudIA [19] is a framework which provides on-demand creation and configuring of VM images so that the students are able to have their own Java servlet environment for experimentation, containing MySQL, Tomcat, PHP, and Apache web server. With this approach, students can focus more on developing, deploying and testing their applications in a servlet container.

In [11], the authors present a new service model that enhances the efficiency within a virtual personalized learning environment. This system is intended for subscribing the selected learning resources as well as creating a personalized virtual classroom, and allows the learning content providers to registry their applications in the server and the learners integrate other internet learning resources to their learning application pools. Other proposals for personal and virtual learning interact with services that rely on the cloud, such as YouTube or GoogleDocs [1].

Finally, we may find some cloud-related works for performing a comparison on the efficiency of online models versus traditional models [7]. The most representative work is this area is developed in [20], where the authors focused on the impact of supporting technologies or the perceived ease of use and acceleration of the learning process. Furthermore, they analyze the appropriate level of abstraction (i.e., IaaS or PaaS) that should be delivered to students to enable them to focus on the course topics.

4 Concluding Remarks

In this work we have exposed the main components of e-Learning, focusing on the flexibility, convenience, easy accessibility, consistency and repeatability of this kind of systems. In this manner, an E-learning system is facing challenges of optimizing large-scale resource management and provisioning, according to the huge growth of users, services, education contents and media resources. We have settle the goodness of a Cloud Computing solution.

The features of the Cloud Computing platform are quite appropriate for the migration of this learning system, so that we can fully exploit the possibilities offered by the creation of an efficient learning environment that offers personalized contents and easy adaptation to the current education model. Specifically, the benefits considering the integration of an e-Learning system into the cloud can be highlighted as good flexibility and scalability for the resources, including storage, computational requirements and network access; together with a lower cost considering the pay-per-use billing format and the save in new hardware and machines and software licences for educational programs.

Finally, we have enumerated several approaches that have been already proposed for addressing e-Learning on Cloud Computing, describing these models and how they take advantage of this environment to enhance the features of the educational system. However, we must stress that these are just initial steps towards an open line for research and exploitation of e-learning and cloud computing platforms.

References

1. Al-Zoube, M.: E-learning on the cloud. Intl. Arab Journal of e-Technology 1(2), 58–64 (2009)
2. Al-Zoube, M., El-Seoud, S.A., Wyne, M.F.: Cloud computing based e-learning system. Intl. Arab Journal of e-Technology 8(2), 58–71 (2010)
3. Buyya, R., Broberg, J., Goscinsky, A.: Cloud Computing: Principles and Paradigms. John Wiley and Sons (2011)
4. Dong, B., Zheng, Q., Qiao, M., Shu, J., Yang, J.: BlueSky Cloud Framework: An E-Learning Framework Embracing Cloud Computing. In: Jaatun, M.G., Zhao, G., Rong, C. (eds.) Cloud Computing. LNCS, vol. 5931, pp. 577–582. Springer, Heidelberg (2009)
5. Duer, W.: CRM, Customer Relationship Management. MP editions (2003)
6. Ercan, T.: Effective use of cloud computing in educational institutions. Procedia - Social and Behavioral Sciences 2(2), 938–942 (2010)
7. Hu, Z., Zhang, S.: Blended/hybrid course design in active learning cloud at south dakota state university. In: 2nd ICETC, vol. 1, pp. V1-63–V1-67 (2010)
8. Hurwitz, J., Bloor, R., Kaufman, M., Halper, F.: Cloud Computing for Dummies. Wiley (2010)
9. Jolliffe, A., Ritter, J., Stevens, D.: The online learning handbook: Developing and using Web-based learning. Kogan Page, London (2001)
10. Kwan, R., Fox, R., Chan, F., Tsang, P.: Enhancing Learning Through Technology: Research on Emerging Technologies and Pedagogies. World Scientific (2008)
11. Liang, P.-H., Yang, J.-M.: Virtual Personalized Learning Environment (VPLE) on the Cloud. In: Gong, Z., Luo, X., Chen, J., Lei, J., Wang, F.L. (eds.) WISM 2011, Part II. LNCS, vol. 6988, pp. 403–411. Springer, Heidelberg (2011)
12. Masud, A.H., Huang, X.: ESaaS: A New Education Software Model in E-learning Systems. In: Zhu, M. (ed.) ICCIC 2011, Part V. CCIS, vol. 235, pp. 468–475. Springer, Heidelberg (2011)
13. Mayer, R., Clark, R.: E-Learning and the Science of Instruction: Proven Guidelines for Consumers and Designers of Multimedia Learning, 3rd edn. Pfeiffer (2011)
14. Ouf, S., Nasr, M.: Business intelligence in the cloud. In: IEEE 3rd International Conference on Communication Software and Networks (ICCSN 2011), pp. 650–655 (2011)
15. Papazoglou, M., Van Den Heuvel, W.J.: Service oriented architectures: Approaches, technologies and research issues. VLDB Journal 16(3), 389–415 (2007)
16. Pocatilu, P., Alecu, F., Vetrici, M.: Measuring the efficiency of cloud computing for e-learning systems. W. Trans. on Comp. 9, 42–51 (2010)
17. Romero, C., Ventura, S.: Educational data mining: A review of the state of the art. IEEE Transactions on Systems, Man, and Cybernetics–Part C: Applications and Reviews 40(6), 601–618 (2010)

18. Sosinksy, B.: Cloud Computing Bible. John Wiley and Sons (2011)
19. Sulistio, A., Reich, C., Doelitzscher, F.: Cloud Infrastructure & Applications – CloudIA. In: Jaatun, M.G., Zhao, G., Rong, C. (eds.) Cloud Computing. LNCS, vol. 5931, pp. 583–588. Springer, Heidelberg (2009)
20. Vaquero, L.M.: Educloud: Paas versus iaas cloud usage for an advanced computer science course. IEEE Transactions on Education 54(4), 590–598 (2011)
21. Vouk, M., Averitt, S., Bugaev, M., Kurth, A., Peeler, A., Shaffer, H., Sills, E., Stein, S., Thompson, J.: Powered by VCL - using virtual computing laboratory (VCL) technology to power cloud computing. In: 2nd Intl. Conference on the Virtual Computing Initiative (ICVCI), Research Triangle Park, North Carolina, USA (2008)
22. Wheeler, B., Waggener, S.: Above-campus services: Shaping the promise of cloud computing for higher education. EDUCAUSE Review 44(6), 52–67 (2009)
23. Witten, I., Frank, E., Hall, M.: Data Mining. Practical Machine Learning Tools and Techniques, 3rd edn. Morgan Kaufmann (2011)

Retrieving Learning Resources over the Cloud

Fernando De la Prieta[1], Juan F. De Paz Santana[1], Ana B. Gil[1], and Magali Boureux[2]

[1] Department of Computer Science and Automation, University of Salamanca
Plaza de la merced s/n, 37008 – Salamanca, Spain

[2] Dip. di Filosofia Pedagogia e Psicologia
Università degli Studi di Verona, Lungadige Porta Vittoria 17, 37129 Verona, Italy
{fer,fcofds,abg}@usal.es, magali.boureux@univr.it

Abstract. Reusing resources has been on the rise recently in the ICT sector. In fact, this trend is expanding into other areas such as the educational sector. Learning objects have made it possible to create digital resources that can be reused in various didactic units. These resources are stored in repositories, and thus require a search process that allows them to be located and retrieved. The present study proposes the AIREH tool, which was deployed into a cloud environment and facilitates the retrieval of learning objects by integrating virtual organizations and agents with CBR systems that implement collaborative filtering techniques.

Keywords: Learning objects, e-learning, cloud computing, collaborative filtering.

1 Introduction

Recent years have seen the rapid evolution, essentially a revolution, in methods for creating, updating and packaging digital resources oriented to education [12]. As a result, it is now possible to create new educational experiences by developing self-contained educational units. Each of these modular components is known as a Learning Object (LO) [9]. It is necessary to create systems and procedures for storing and retrieving LOs in a way that allows the content to be easily reused.

A LO can be described as a digital, self-contained and reusable entity with a clear educational purpose, composed of at least three internal components: content, learning activities, and contextualization elements. Additionally, they must include an external structure that facilitates their identification, storage and retrieval in metadata format [5]. The advantages of this new technology seem to be evident [5]: reuse, personalization, durability, granularity, flexibility and accessibility of educational resources, all of which involve twice the savings in cost and time for preparing complete educational activities. However, there are also disadvantages [13], primarily the period of adaption required by educators to a new educational context in which new technologies are coupled with innovative methodologies for creating and publishing didactic resources. This can be considered a new paradigm in the educational sector.

This new paradigm initiates a new set of technological goals related to the new life-cycle of creating educational experiences. Among the most important of these

L. Uden et al. (Eds.): Workshop on LTEC 2012, AISC 173, pp. 47–56.
springerlink.com

are, first of all, the creation of packets with educational content, a situation that seems to have been overcome[12] by the establishment of international standards for LO[7], and other standards such as SCORM (Sharable Content Object Reference Model) [10]. Secondly, the publication of these educational resources, which is being achieved through LO Repositories (LOR). Despite their overly heterogeneous technology [14], LORs make it possible to access resources through standardized query languages [8]. Finally, the need for tools that can quickly, simply and efficiently search and recover LOs from repositories.

This study presents the AIREH tool (*Architecture for Intelligent Recovery of Educational content in Heterogeneous Environments*) [18], which makes it possible to search and recover educational resources encapsulated in the form of a LO. Similarly, a system can use a CBR (Case-Based Reasoning) system to recommend which educational resources might be of particular interest to the user, based on information from previous searches. This system is based on Multi-Agent Systems (MAS) based on virtual organizations (VO). Finally, it should be noted that this application was deployed in a Cloud Computing environment, which allows users to store information about the recovered resources in a cloud.

This study is organized as follows, next section establishes the state of the and the related work, section 3 shows the proposal system, section 4 the experiments and the case study and the last sections contains the conclusions an future work.

2 State of the Art and Related Work

2.1 Educational Technology

The concept of LO has become a central component within the new technological paradigm in an educational context. Revolving around the concept of LO are such relevant elements as LMS (Learning Management System), LOR, authoring tools, or systems for discovering didactic resources, among others. The IEEE's Learning Technology Standards Committee (LTSC) defines a LO in general terms as *any entity, digital or non-digital, which can be used, re-used or referenced during technology supported learning*. In short, practically any educational resource can be considered a LO, a fact heavily criticized by various authors [5][8][9] who have tried to delimit the concept as much as possible.

What seems perfectly clear is that any LO must be associated with an external structure that facilitates its search, evaluation, recovery and eventual reuse. There are currently different standards and specifications specifically designed to describe educational resources by means through the use of metadata. The most relevant standards are Dublin Core [6], which is more oriented towards digital resources in general and is quite widespread within a library context; IEEE LOM (Learning Object Metadata) [7], which is the most commonly used standard for describing LOs; and finally SCORM [10], which is oriented to packaging and distributing complete educational activities.

LOs are commonly stored in repositories, which are characterized by their heterogeneity [17], including different storage systems, access to objects, query methods, etc. The heterogeneity itself is not a problem, since there are different tools that can isolate the internal logic of the LOR from the exterior, which in fact makes it possible to automatically search different repositories simultaneously using a single query application. Most notable among these tools, which serve as a middleware layer between the repository and the clients, are SQI (Simple query Interface)[15], which was standardized by the ECS (European Committee for Standardization), and OAI-MPH (Open Archives Initiative Protocol for Metadata Harvesting) [16]. The basic function of these systems is, at a conceptual level, trivial, being based on web services, which allows the client to perform a query in a LOR, which in turn provides results in the form of LOs. The importance of this type of query method stems from the fact that it is not limited to a particular query language or format in which the query results are returned, nor are the internal characteristics of the server restricted.

LOR are currently classified according to their topological distribution as follows:

- **Autonomous repositories.** Although the opposite would seem to be true, these are currently the most widespread repositories. They do not have a system that allows an external search, and thus require searches in situ, normally using a web interface.
- **Middleware repositories [11].** They have an external search interface and may include an automated search system.
- **Federated search systems [4].** These systems, including the one used for the present study, perform automatic searches in other repositories, but do not store LOs or educational resources.
- **Repositories with federated search [11].** In addition to performing internal searches, they also perform automated searches in other repositories.

2.2 *Cloud Computing*

Cloud computing uses a new technology that develops applications in a way that allows both the execution of the application and the storage of data to be performed ubiquitously for all users. Cloud architecture provides user support at different levels that vary according to their different characteristics. The different levels can be described as follows [2]:

- Infrastructure as a Service (IaaS): remote management and control of hardware resources provided by a system.
- Platform as a Service (PaaS): offers the cloud platform along with a series of libraries to develop applications in which the distribution of tasks, the persistence and other layers are transparent for the developer.
- Software as a Service (SaaS): consists in offering different applications to be used through the internet as opposed to a local installation.

There are incipient developments that broach the topic of cloud technology and e-learning [1], however few studies incorporate both concepts. As demonstrated in

[1], cloud computing and e-learning are fundamentally centered on the SaaS layer. Other research such as [2] has studied the applicability in different fields such as education, but also focuses on the SaaS layer, ubiquitously offering a set of applications to users. Due to these circumstances, and given the possibilities offered by cloud computing in the field of e-learning for managing applications in the SaaS layer, it has become necessary to develop applications in cloud architectures that can form part of the PaaS layer, and that can manage the LOs that are stored within the architecture, thus facilitating the interaction and access of the applications developed in the system.

2.3 *Related Work*

The primary goal of search systems is, unquestionably, to provide users with the most appropriate LO, not just for a particular search, but for their general interest as well.

This goal has attracted a great deal of research in this field, especially research focused on selecting and recommending LO. The most commonly used recommendation techniques are collaborative filtering and data mining. The majority is based on existing information on educational resources (contact-based approach), on the actual user (collaborative approach), or on a combination of both (hybrid approach).

At present there is no platform that can perform this type of search or provide recommendations for LO, while being deployed within a Cloud Computing environment.

3 Proposal: AIREH

The main objective of the present study is to recover LO from LOR by using federated searches. The search system can also filter and classify data according to a set of rules. These rules are generated according to the metadata that describe LO and provide the information required to offer the most appropriate educational resources to each user. The proposed system is presented as an intermediary point of communication between the LORs, the LOs that they store, and the end users.

Given the highly heterogeneous nature of this context, the use of MAS, which takes social norms and organizations into account, was selected to develop the federated search system. This type of architecture makes it possible to describe the functionality using a set of roles, each of which can be carried out by any agent with the necessary capabilities. In practice, this new concept was critical since the changes that can take place within the context can be easily introduced into the system.

Figure 1 displays a structural diagram of AIREH, and presents a high number of important characteristics (and products): UserUnit, SearchUnit, CataloguingUnit, StatisticsUnit, and the Administrative Unit, each of which follows a congregation type pattern within the organization of intelligent agents. The provider and consumer roles are also presented.

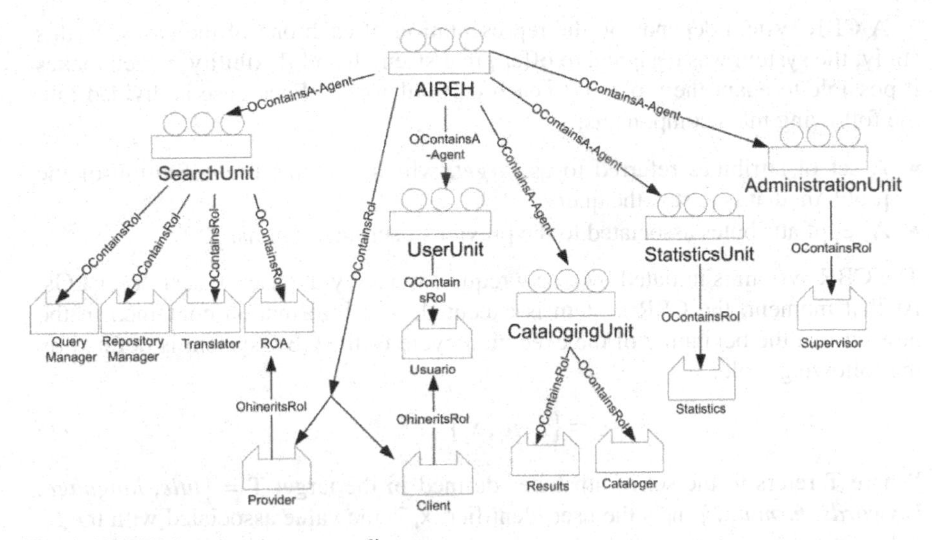

Fig. 1 AIREH functional structure diagram

The complete process comprises three main phases:

1. The selection of the LOR, in which the best repositories are selected according to the statistical parameters that have been gathered from previous queries.
2. Once the LORs have been selected, the next step is to recover the LO according to the search performed by the user. Additionally, it is during this phase that the obtained results are homogenized to facilitate the next phase of the process.
3. The final phase of the process includes a second filtering phase that takes into account the aspects of quality, such as size, completeness, etc., of the metadata. System users also evaluate the objects during this phase, using a voting system in previous searches. The following section provides a more in-depth review of the system.

3.1 Recommedation Strategy

A recommendation system is a tool that predicts user likes according to their characteristics, interests or abilities, based on previously obtained information. There are various techniques based on Artificial Intelligence (AI) which are oriented to carrying out these tasks. One of them is CBR.

The purpose of CBR is to solve new problems by adapting solutions that have been used to solve similar problems in the past [3]. The primary concept when working with CBRs is the concept of case. A case can be defined as a past experience, and is composed of three elements: a problem description which describes the initial problem, a solution which provides the sequence of actions carried out in order to solve the problem, and the final state which describes the state achieved once the solution was applied. A CBR manages cases (past experiences) to solve new problems. The way cases are managed is known as the CBR cycle, and consists of four sequential steps which are recalled every time a problem needs to be solved: retrieve, reuse, revise and retain.

A CBR system depends on the representation of each one of the cases. In this study, the system was designed to offer great strength and flexibility, which makes it possible to adapt the problem to each particular case. Each case is divided into the following main components:

- A set of attributes referred to as *target*, which contains the definition of the problem, that is to say, the query.
- A set of attributes associated to the previous user interactions.

The CBR system is initiated by a new request made by the user to search for LOs. At that moment, the CBR system is executed. The information contained in the new case at the beginning of the execution cycle of the CBR system is defined by the following tuple:

$$c = \{T, u_i, x_i\} \quad (1)$$

Where T refers to the set of attributes defined in the target T = {*title*, *language*, *keywords*, *format*,...}, u_i is the user identifier, x_i is the value associated with the final solution.

Using the information defined in (1), the reasoning cycle for the CBR system is initiated. Figure 2 illustrates the reasoning cycle. During the retrieve phase the metadata for the learning objects are downloaded from different repositories using simultaneous searches. The information related to the recovered objects is recovered from the cloud. The Slope One method is applied during the reuse phase in order to predict the degree of relevance of the recovered LO. Finally, during the revise and learning phase, information related to the user's final assessment is stored. The following section explains the different steps for the reasoning cycle in greater detail.

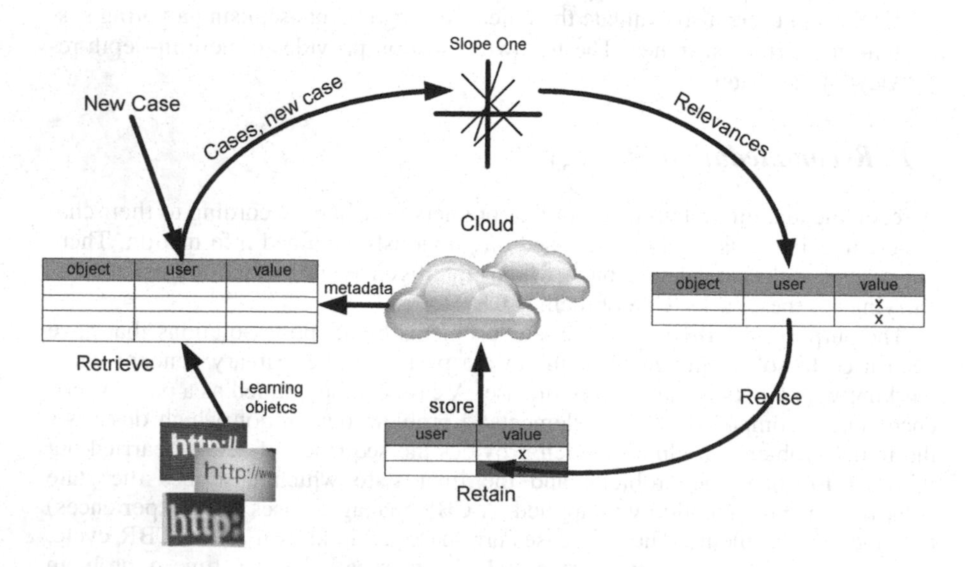

Fig. 2 CBR system implemented in AIREH

3.1.1 Retrieve

During this phase, the LOs are downloaded from the repositories according to the indicated targets. Relative to the LOs, existing information for the stored objects is then retrieved from the cloud. Once the information has been recovered from the repositories and the cloud environment, different cases are obtained according to the structure indicated in (1).

3.1.2 Reuse

The information listed in table 1 is obtained from the data found during the retrieve phase. Each cell contains a value v_{ij} that represents the user's evaluation of the learning object.

Table 1 Information retrieved from the cases.

	LO_1	LO_2	LO_3	...	LO_m
u_1	v_{11}	v_{12}	v_{13}	...	v_{1m}
u_2	v_{21}	--	v_{23}	...	v_{2m}
u_3	v_{31}	v_{32}	--	...	v_{3m}
...				...	
u_n	v_{n1}	v_{n2}	v_{n3}	...	v_{nm}

The average is calculated for each pair of individuals (2). The final averaged values could be combined according to (3), with a weighted average relative to the number of predictions that exist for each article.

$$\bar{d}_{ij} = \frac{\sum_{k=1}^{m-1}(v_{ik} - v_{jk})}{m-1} \tag{2}$$

Where v_{ik} represents individual i for which the unknown value is being calculated, m is the number of values that exist for both articles i and j (if v_{ik} is unknown, v_{jk} will not be considered in the calculation), v_{jk} is individual j.

$$x_{ik} = \frac{\sum_{j=1}^{n-1} m_j \bar{d}_{ij}}{\sum_{j=1}^{n-1} m_j} \tag{3}$$

Where v_{ik} represents individual i for which the unknown variable for k is calculated, m_j is the number of values that exist for category j, v_{jk} is individual j.

3.1.3 Revise and Retain

During the revise and retain phase, the user rates the objects retrieved during the reuse phase. The values are then stored in the cloud for future retrievals.

4 Experiments and Conclusions

The Merlot[1] and Lornet[2] repositories were used to carry out test son the LO. These repositories were used to perform searches by 40 users on a group of 60 different key words, taken from the UNESCO code. Each user input a key word and then analyzed the predictions made for the previous 15 predictions. The values were assigned to each item on a scale of 1 to 5. The implementation of the algorithms was based on the Apache Mahout library, which provides techniques such as Map Reduce, allowing a high level of efficiency in multiprocessing systems.

The first step was to compare the execution times for different alternatives to collaborative filtering in order to determine the viability of the different solutions. The execution times were based on simulated data, starting with the first test of 500,000 pieces of data and a second of 5,000,000. Table 2 lists the calculation times to obtain the recommendations.

Table 2 Information retrieved from the cases

Elements	KNN	Slope One	SVD
500.000	43s	39s	38s
5.000.000	6:37s	5:36s	5:52s

In order to analyze the efficiency of the CBR system, the predictions were compared with other methods of collaborative filtering. The techniques selected were KNN (K-Nearest Neighbour) and SVD (Single Value Decomposition).

While the different times for constructing the recommendations are very similar, the difference is due to the fact that the KNN algorithm needs the same execution time for any prediction made for a different user, while the Slope One and SVD have a prediction time for execution of less than one second, regardless of the user. The results shown in Table 3 indicate the average error values obtained by the methods indicated in each column. The weighted values are based on a scale of 1 to 5.

Table 3 Information retrieved from the cases

KNN	Slope One	SVD
1.30	0,76	0.78

---

[1] MERLOT, Multimedia Educational Resource for Learning and Online Teachning (http://merlot.org)

[2] LORNET, Learning Object Repository Networks (http://www.lornet.com).

The results in Table 2 indicate that Slope One provided the best results, although very similar to those obtained by SVD. The reason for not using SVD is that it is necessary to determine statistically the number of elements that reduce the dimensionality, which would involve the analysis of the value with subsequent executions.

The system is still in a process of development and undergoing more detailed testing, which will allow for more extensive results in the future. With AIREH it is possible for the user to retrieve LO efficiently and simply, since it allows the retrieved elements to be filtered according to each user and their previous actions.

Acknowledgments. This work has been partially supported by the MICINN project TIN 2009-13839-C03-03.

References

[1] Ercan, T.: Effective use of cloud computing in educational institutions. Procedia Social and Behavioral Sciences 2, 938–942 (2010)

[2] Sultan, N.: Cloud computing for education: A new dawn? International Journal of Information Management 30(2), 109–116 (2010)

[3] Kolodner, J.: Case-Based Reasoning. Morgan Kaufmann (1993)

[4] De la Prieta, F., Gil, A.: A Multi-agent System that Searches for Learning Objects in Heterogeneous repositories. In: Trends and Strategies on Agents and Multiagent Systems: 8th International Conference on Practical Applications of Agents and Multiagent Systems, pp. 355–362 (2010) ISSN 1615-3871 (Print) 1860-0794 (Online)

[5] Chiappe, A., Segovia, Y., Rincon, H.Y.: Toward an instructional design model based on learning objects. Educational Technology Research and Development 55, 671–681 (2007)

[6] Dublin Core Metadata Initiative (DCMI), `http://dublincore.org` (accessed 2007)

[7] IEEE 1484.12.1-2002, Draft Standard for Learning Object Metadata. The Institute of Electrical and Electronics Engineers, Inc.

[8] Simon, B., Massart, D., Van Assche, F., Ternier, S., Duval, E., Brantner, S., Olmedilla, D., Miklos, Z.: A Simple Query Interface for Interoperable Learning Repositories. In: 1st Workshop on Interoperability of Web-based Educational Systems, Chiba, Japan (2005)

[9] Lujara, S.K., Kissaka, M.M., Bhalaluseca, E.P., Trojer, L.: Learning Objects: A new paradigm for e-learning resource development for secondary schools in Tanzania. In: World Academy or Science, Engineering and Technology, pp. 102–106 (2007)

[10] SCORM 2004, 4th edn. (March 2009)

[11] Ternier, S., Verbert, K., Parra, G., Vandeputte, B., Klerkx, J., Duval, X., Ordóñez, V., Ochoa, X.: The Ariadne Infrastructure for Managing and Storing Metadata. IEEE Internet Computing 13(4), 18–25 (2009)

[12] Wiley, D.A.: Connecting learning objects to instructional design theory: A definition a metaphor, and a taxonomy. In: Wiley, D.A. (ed.) The Instructional Use of Learning Objects, IN, Association for Educational Communications and Technology, Bloomington (2001), `http://www.reusability.org/read/`
(retrieved March 23, 2009, from the World Wide Web)

[13] Polsani, P.: Use and Abuse of Reusable Learning Objects. Journal of Digital Information 3(4), arto 164 (2003)
[14] Hatala, M., Richards, G., Eap, T., Willms, J.: The Interoperability of Learning Object Repositories and Services: Standards, Implementations and Lessons Learned. In: 13th World Wide Web Conference, Educational Track, New York, pp. 19–27 (May 2004)
[15] European Committe for Standardization, A Simple Query Interface Specification for Learning Repositories (November 2005)
[16] Lagoze, C., Van De Sompel, H., Nelson, M., Warner, S.: The Open Archives Initiative Protocol for Metadata Harvesting. Open Archive Initiative. Version 2.2 (2002)
[17] De la Prieta, F., Gil, A.B.: A Multi-agent System that Searches for Learning Objects in Heterogeneous Repositories. In: Demazeau, Y., Dignum, F., Corchado, J.M., Bajo, J., Corchuelo, R., Corchado, E., Fernández-Riverola, F., Julián, V.J., Pawlewski, P., Campbell, A. (eds.) Trends in PAAMS. AISC, vol. 71, pp. 355–362. Springer, Heidelberg (2010)
[18] Gil, A.B., De la Prieta, F., Rodríguez, S.: Automatic Learning Object Extraction and Classification in Heterogeneous Environments. In: Pérez, J.B., Corchado, J.M., Moreno, M.N., Julián, V., Mathieu, P., Canada-Bago, J., Ortega, A., Caballero, A.F. (eds.) Highlights in Practical Applications of Agents and Multiagent Systems. AISC, vol. 89, pp. 109–116. Springer, Heidelberg (2011)

Accounting Students Will Live in the Cloud

D'Arcy Becker and Dawna Drum

University of Wisconsin – Eau Claire, Eau Claire, Wisconsin, USA
{dbecker,drumdm}@uwec.edu

Abstract. Cloud computing is growing and its implications for accounting and auditing are unlimited. As familiarity among risk averse and technology-challenged accountants morphs into a mature understanding of the benefits of the cloud, use of the cloud by accountants will be as ubiquitous as use of the computer itself. This paper addresses areas of training accounting students need to understand so they will be prepared for use of the cloud by their employers, and so they can help their employers move ahead with the cloud in a controlled, safe manner.

Keywords: cloud computing, accounting, audit firms, internal control.

1 Introduction

The term 'Cloud Computing' describes a variety of tools that provide completely outsourced Internet-based solutions to the need to access applications and information anywhere anytime (Kepcyk, 2011). Cloud computing is a new method of delivering computing resources, not a new technology (Julisch and Hall, 2010).

Cloud computing is possible due to the high-speed internet connections available to most organizations, and can apply to both hardware and software; organizations can use these resources as needed, rather than purchasing them outright (Gezcy, Hazuma and Hasida, 2012).

There are simple examples such as Dropbox.com, which allows users to store data on the Dropbox.com website and access it from anywhere for free as long as the user stays under a pre-defined size limit. There are also full-fledged Enterprise Systems that encompass every aspect of an organization's software needs such as Netsuite.com. One common accounting cloud application is CCH Research. As with all cloud applications, the vendor takes full responsibility for maintenance and infrastructure.

Accounting firms are moving increasingly toward a comprehensive set of applications in the cloud, encompassing practice management, document management, billing, workflow and audit tools (Kepcyk, 2011). Cloud computing creates unprecedented opportunities for CPA firms to work faster and communicate with clients better (Drew, 2012).

The cloud removes infrastructure and capital expense as barriers to entry for businesses of all types, and allows startups to scale up cheaply and rapidly. In addition, it gives small accounting firms and their clients access to information security they otherwise would have trouble affording (Drew, 2012).

L. Uden et al. (Eds.): Workshop on LTEC 2012, AISC 173, pp. 57–63.
springerlink.com

Public accounting has many potential different uses of cloud computing, but most fall into the category Software as a Service (SaaS). SaaS is application software that is hosted by third parties and provided as a service over the Internet (Julisch and Hall, 2010). Firms can integrate statistical sampling, data mining projects, accounts receivable confirmations and many other tasks with practice management, time tracking, client billing and payroll applications. All of this can be done without buying software that will soon be outdated, or may not seamlessly integrate within the firm's existing technology infrastructure.

A second common category of cloud computing in accounting is file storage. Firms may ask clients to upload their data to the cloud to allow accountants to work with the data; firms may have geographically-dispersed audit teams communicate audit outcomes and evidence to each other via the cloud; firms may store records on the cloud rather than in their own file rooms or on their own servers (Drew, 2012). At the limit, accounting firms need only to have cheap computers with an Internet connection to manage clients, find new sources of revenue and expand operations.

Business people do not necessarily have a full understanding of risk and control prior to implementing technology changes. For example, a recent survey showed that that one in five executives has bought a cloud service without the knowledge of their information technology (IT) department (Steffee, 2011). This implies a great need to ensure future accountants understand and effectively convey the need for controls and governance of activities in the cloud.

2 Moving Accounting Firms to the Cloud

Accountants who choose to use the cloud have to understand two very different sets of issues as the use of cloud computing expands: risks related to the firm's use of the cloud for its own work, and risks and controls when clients provide information to the firm's cloud. . Issues in these areas are discussed in this section.

2.1 Accounting Firms Choose to Use the Cloud

The cloud's essential characteristics are on-demand self-service, broad network access (accessible by any networked device), resource pooling (allowing multiple clients to use the same resources regardless of location), rapid elasticity (allowing quick changes as demand on the system changes), and measured service (monitored, controlled, reported, and billed for) (ISACA, 2009).

Accounting firms are able to benefit from each of these characteristics. Auditors working at clients need access to many different applications on demand. They may need audit workflow, the audit program, client billing, accounting pronouncements, or tax laws, among others, in the course of a single day. Further, the auditor may need this access at the firm's office, the client's locations, their hotel or their personal home. In addition, which accounting firm personnel require which applications and which data can change during the course of an accounting engagement, making rapid scalability essential.

Accountings firms increasingly realize that the cloud provides a solution to many of their information access issues. Kepcyk (2011) discusses alternatives arrangements accounting firms use in the cloud, stating that accounting firms are a perfect candidate for cloud computing because they need to have integrated use of many applications from many vendors. When applications cannot integrate on a real-time basis, accounting engagements may experience workflow disruptions that cause ineffective and inefficient use of accountants' time.

Some firms have developed a 'firm cloud' or virtual private network. The biggest challenge with VPNs developed by large firms is to provide sufficient bandwidth to allow the network to function in a reasonably efficient manner.

An alternative choice for some firms is 'thin' client technology such as Citrix or Windows Terminal Server, which expands network access to users with SmartPhones and similar technology. This solves the bandwidth problem, and increases the likelihood that accountants working from rural locations can access the firm's network. It also removes the need for the client to provide the firm with secure bandwidth during accounting firm fieldwork.

One example of a firm that has taken full advantage of the cloud is Blumer & Associates, CPAs, PC (www.BlumerCPAs.com). Owner Jason Blumer saw that the cloud would allow his practice to be more agile, work from anywhere and serve clients anywhere in the world. Based in South Carolina, U.S., the firm serves clients across the U.S. and internationally (O'Bannon, 2010).

Another example is RSM McGladrey Consulting. They found a major advantage of the cloud was its ability to allow the firm to use resources from any location to serve clients in any location. They also found that the cloud allowed them to eliminate duplicate processes across locations, increasing efficiency (Drew, 2012).

Awad (2011) points to two important considerations before moving to a cloud environment: compatibility with various applications and security. Not all applications work well in the cloud; a lot depends upon a user's specific application needs. If a firm uses applications not specifically designed for the cloud they may find functionality is compromised.

Security presents a more difficult challenge. Firms must keep their clients' data secure. Beyond that, the firm's proprietary analysis methods and audit steps, its decision making and billing among other areas must be safeguarded. Security in a VPN environment is not the same as security in an Internet environment. Whether the data is in the accounting firm's offices or in the cloud, loss of that data is not a problem accounting firms can shift to SaaS providers or their clients (Julisch and Hall, 2010).

How can our students be prepared to assist their firms with this transition? Moving an organization to cloud computing bears many similarities to more traditional system implementations, such as training, quality assurance and vendor relations. There are case studies and simulations to help students grasp these concepts in general terms. Critical analysis of available vendor privacy statements and contracts such as Amazon's Web Services or Google's Apps for Business would be beneficial, as would interviews with organizations who have implemented some form of cloud computing.

There are many avenues for students to experience working in the cloud, from simple storage to complex homework assignments in cloud-based enterprise applications. Many software companies are pleased to provide extensive access and materials for instructors and students, as they understand the need for future knowledge workers to understand these complex topics. (uac.sap.com)

2.2 *Helping Clients Move to the Cloud*

While accounting firms are increasingly interested in cloud-based solutions, the cloud concept only works when both the firm and its clients use the cloud. It will be interesting to see whether use of the cloud in accounting expands because accounting firms pull their clients along, or whether the impetus for using the cloud comes from the clients.

To date, cloud computing adoption by organizations has been minor despite the initial optimism. The primary concerns obstructing adoption of cloud-based services are security and loss of control over data (Geczy, Izumi, and Hasida, 2012). The Ernst & Young (2009) survey found that 56% of companies realize that outsourcing information management to the cloud can help address those concerns. However, the presumption that outsourcing is too expensive continues to have traction in inhibiting movement to the cloud.

Many clients who are trying to move away from legacy applications with specialized data structures and associated databases are attracted to the cloud (Blumenthal, 2011). Software providers are developing cloud-based alternatives for all sizes and types of organizations. Even SAP, long thought to be a behemoth software package suitable only for the largest organizations, has developed a new solution called "Business by Design," targeted at small enterprises (SAP, 2012).

Once clients are willing to consider moving to the cloud, future accountants will need to understand the breadth of parameters that go into a negotiation between an accounting firm and its clients when the firm implements a cloud system. What kinds of access will clients have to their cloud applications? Is that access limited in nature, is it always available? How much can it be customized? How can clients be assured that their data are confidential and safe? What forms can the data take? Will the cloud integrate with the client's other applications? If the cloud applications change, will the client be able to adjust?

Further, using cloud computing to expand the range of services for which clients are billed has its own perils. Firms must carefully identify the risks they assume when making these arrangements. Clients may have to meet regulatory reporting requirements; failure of the cloud at an inopportune time would put the firm at high risk if client reporting cannot be accomplished on a timely basis.

Because this area is so new to both clients and accountants, controls in this environment are not as well understood and thus users may be exposed to higher risks. Reporting standards are changing to keep pace with the rapid changes, and SAS70 was superseded in June, 2011, by SSAE16 to accommodate service organizations such as cloud providers. (Du and Cong, 2010).

The American Institute of Certified Public Accountants (AICPA) has subsequently provided guidance in the form of service organization control (SOC) reports to aid users in assessing the risks associated with outsourcing portions of their operations (Rashty, 2011). Accountants must become familiar with these reports and their implications in order to provide their clients with sound advice.

Additionally, traditional firms that use the cloud on a limited basis will have to compete with accounting practices that exist only in the cloud. DeepSky (www.deepsky.com) is a cloud-only provider of accounting services including such mundane areas as payroll, invoicing and general bookkeeping as well as overall business consulting (Fineberg, 2011). Chaney & Associates (www.chaneyassociates.com) is a cloud-only provider of controller services (Drew, 2012). Online-only competition exists in all fields of business, but online-only general accounting, designed to replace the company's entire accounting department, is new.

The field of accounting has experience with outsourcing data entry and simple data manipulations, but outsourcing the entire function is much more bold. From a student training perspective, these new cloud-based business models present a huge set of opportunities. We can train students to build such practices, to work at them, to understand them enough to buy their services, to make smart decisions about the role of this business model in the field of accounting, and so forth.

2.3 *Internal Control and Governance in the Cloud*

Accounting firms need to ensure clients understand how the cloud environment differs from other data storage environments. SaaS can safeguard against disruptions caused by weather, strikes and ill health (Meall, 2011), and has the potential to provide a more secure data environment than an accounting firm could otherwise provide.

However, this is not automatically the case. The client must implement suitable controls to prevent threats such as the Web browser vulnerabilities, theft of authentication credentials, virus attacks, or data theft by rogue employees (Julisch and Hall, 2010).

In fact, Farrell (2010) points out that governance, risk and compliance issues are the dark side of SaaS. Ernst & Young (2009) found that 41% of respondents across industries reported an increase of external related security attacks, and 25% reported increased internal security attacks.

These incidents can have serious impacts both internally as well as financially across the cloud computing spectrum, as seen in the recent cases of Sony and Amazon. Sony provides cloud access for its Playstation game system, and in 2011 this was hacked, exposing the data of over 100 million customers. Amazon offers data storage on large scales for customers, and these customers were unable to access that data for an extended period.

These highly visible failures of cloud computing services caused not only the problems we would expect such as lost customers, financial losses, etc., but they additionally caused a ripple effect in the cloud computing industry as customers reconsidered their choices. SalesForce.com, a major provider of cloud software for sales organizations, saw its stock drop 3% in the weeks following these incidents, and VMWare, a provide of software for building cloud services, saw its stock drop 2% (Finkle and Baker, 2011).

Further, the laws of a country impact online information storage. The vendor's location determines which laws regarding data privacy and protection apply (McClure, 2011). Adding to the challenge is the difficulty in determining where the vendor is located. Many providers of cloud services have very limited physical presence.

3 Conclusion

Accounting educators can lead the way to help future accountants understand the potential uses and risks of the cloud. Perhaps the most valuable skill to be gained is the ability to critically analyze emerging technologies of any sort. Vasarhelyi, Teeter and Krahel (2010) proposed that there are three areas that educators should address when preparing accountants for a future which will be heavy in technology use. These are attitudes, behaviors, and objective knowledge.

An important component of "attitude" is the understanding that technology will be learned on the job, and therefore the skill to be cultivated is rapid assimilation of new technologies rather than expertise in any particular technology. A positive attitude toward emerging technologies seems to be more evident in the current generation of college students than it has been in the past.

"Behavior" can be encouraged by giving students resources and tutorials, rather than giving detailed directions on the use of a particular technology. When students are taught to use technology by following prescribed clicks and entries, they may not be prepared to learn new technologies without that level of support.

"Objective knowledge" ensures that students understand that they need to go beyond the basic office applications and understand the deeper workings of information systems. The ability to research problems and constantly seek new tools and information will better equip them to cope with the fast pace of technology change and the variety of applications they will see in their own work and at client sites.

References

Awad, R.: Considerations on Cloud Computing for CPAs. CPA Journal 81(9), 11 (2011)

Blumenthal, M.S.: Is Security Lost in the Clouds? Communications and Strategies (81), 69–86 (2011)

Drew, J.: Heads in the Clouds: Part 1. Journal of Accountancy 213(2), 20–23 (2012)

Du, H., Cong, Y.: Cloud Computing, Accounting, Auditing, and Beyond. The CPA Journal 80(10), 66–70 (2010)

Ernst, Young: Outpacing Change. 12th Annual Global Information Survey (2009), http://www.ey.com/Publication/wwLUAssets/12th_Annual_GISS/FILE/12th_Annual_GISS.pdf

Farrell, R.: Securing the Cloud-Governance, Risk, and Compliance Issues Reign Supreme. Information Security Journal: A Global Perspective 19(6), 310–319 (2010)

Fineberg, S.: A New Firm Paradigm. Accounting Today 25(6), 10–11 (2011)

Root, D.: Client Accounting Online... Are You There Yet? CPA Technology Advisor 20(8), 26 (2010)

Finkle, J., Baker, L.: Analysis: Sony Woes may cause some to Rethink Cloud Computing, Reuters, http://uk.reuters.com/article/2011/05/06/us-sony-cloud-idUKTRE7455C020110506 (retrieved March 29, 2012)

Geczy, P., Izumi, N., Hasida, K.: Cloudsourcing: Managing Cloud Adoption. Global Journal of Business Research 6(2), 57–70 (2012)

ISACA. Cloud computing: Business benefits with security, governance and assurance perspectives, http://www.isaca.org/Knowledge-Center/Research/ResearchDeliverables/Pages/Cloud-Computing-Business-Benefits-With-Security-Governance-and-Assurance-Perspective.aspx (2009)

Julisch, K., Hall, M.: Security and Control in the Cloud. Information Security Journal: A Global Perspective 19(6), 299–309 (2010)

Kepczyk, R.: Remote Access "Cloud" Hosting Options for CPA Firms. CPA Practice Management Forum 7(5), 10–11, 13 (2011)

McClure, D.: Backup on Tap. Accounting Today 25(8), 50–51 (2011)

Meall, L.: Split Personality. Accountancy 147(1413), 53–54 (2011)

O'Bannon, I.M.: The 'Certifiable' Public Accountant. Technology 20(7), 28 (2010)

Rashty, J.: New Guidance for Cloud-Based Service Control Reports. The CPA Journal 81(10), 68–71 (2011)

SAP, http://www.sap.com/solutions/technology/cloud/business-by-design/highlights/index.epx

Steffee, S.: Cloud computing governance remains elusive. Internal Auditor 68(5), 14 (2011)

Vasarhelyi, M.A., Teeter, R.A., Krahel, J.P.: Audit Education and the Real-Time Economy. Issues in Accounting Education 25(3), 405–423 (2010)

A Blended Learning Model for "Multimedia Systems" Course

Natasa Hoic-Bozic, Martina Holenko Dlab, and Ema Kusen

University of Rijeka, Department of Informatics, Omladinska 14, Rijeka, Croatia

Abstract. This paper describes the e-learning model used in a course "Multimedia Systems," given at the University of Rijeka, Croatia. The course was taught in a blended way, combining self-paced learning, f2f classroom learning, and online learning supported by the Moodle learning management system. The paper describes the technology for, and the methodological approach to, course design and development, as well as the results of evaluation. A survey conducted in the end of the course showed that students were satisfied with the new blended learning approach for the course. The quality of this course was also recognized by the e-learning experts: in the academic year 2010/2011 the "Multimedia systems" was chosen as the best e-course on the University being given the highest mark according to the four elements of a review.

1 Introduction

The University of Rijeka was the first university in Croatia that brought the "Strategies of implementation of e-learning" in 2006 after noticing how much e-learning contributes to the quality of education in the Bologna model, or more successfully achieves the learning outcomes and creates the conditions for optimal professional and personal development of a student [3, 4, 8].

According to the strategy, "e-learning is a term that describes an educational process promoted by the use of new information and communications technologies (ICT)" [3]. It is about any form of learning, teaching or educating that is supported by primarily those technologies that are based on the Internet (Web). Furthermore, it is emphasized that apart from technology, special attention should be given to pedagogic-methodical aspects and the chosen course strategies should motivate students, promote deep-level learning, encourage communication between instructors and students, ensure feedback and give continuous support during the learning process.

Apart from the ensured and needed technological infrastructure and support (the Moodle LMS named MudRi, courses for instructors) on the University, other measures of encouragement were taken to introduce the elements of e-learning, especially blended learning, to courses. One of the most attractive measures is the award for the best e-learning course or e-course on the University.

In the academic year 2010/2011 the course "Multimedia systems" (MMS) at the undergraduate study of Computer Science was chosen as the best e-course on

L. Uden et al. (Eds.): Workshop on LTEC 2012, AISC 173, pp. 65–75.
springerlink.com

the University being given the highest mark according to the four elements of a review. The course was evaluated by three reviewers and a proof-reader.

The reviewers had a task to assess the *elements of the content* (technical and scientific correctness of the content, relevance of the content in technical and scientific terms, purpose of the used content for achieving the learning outcomes, compliance with the rules of intellectual property and copyright), *elements of the instructional design* (implementation, quality and technical validity of the used technology, implementation and quality of basic elements of the e-course, implementation and quality of the multimedia elements, graphic design and the design of the user-interface, clarity, equalisation and effectiveness of the navigation) and *methodical and didactical elements* (clarity of reported learning outcomes and teaching, compatibility of the learning outcomes and the approach of studying and teaching, efficiency of the chosen methods/activities in e-learning, encouraging the communication between students in the e-environment, a relationship between the learning outcomes and the way of evaluating the knowledge).

In this paper we will present the course MMS and the used blended e-learning model, as well as results of an evaluation of the student's knowledge and the results of a survey. Undergraduate students at the University of Rijeka are mostly enrolled in the courses that use the LMS as a repository of the digital content, without the proper implementation of pedagogical aspects of e-learning. Our model contributes to the promotion of development of e-courses that meet the pedagogic-methodical guidelines stated in the University strategy. The blended model is suitable for solving the mentioned problem since it, besides the inclusion of new technological features, takes into account the pedagogical principles [7, 8]. In addition, it encourages communication between instructors and students, which is important to overcome the lack of personal contact [1].

The remainder of this paper is organized as follows: Section 2 describes the academic context of the course, learning objectives, approach to blended learning, and learning activities, Section 3 introduces the technology used, Section 4 presents the evaluation results, and finally Section 5 presents conclusions and some future plans.

2 Course Design and Development

2.1 Academic Context

The course "Multimedia Systems" was designed for students of the undergraduate program in Computer Science at the Department of Informatics, University of Rijeka. It is taught in the summer semester, third year of studies, with two hours of lectures and two hours of practical exercises per week. The course was given 5 ECTS points.

The course was taught in a blended way, combining self-paced learning, f2f classroom learning, and e-learning supported by the Moodle LMS. Students were directed to use the tools of the mentioned system since they enrolled into the course. It was pointed out that they would not be able to achieve the desired learning outcomes for this course unless they start using the system MudRI from beginning.

Theoretical part of the course for which the classic lectures were predicted, was fully adjusted for the online teaching because the content was popular, attractive and not too demanding for the students, as well as known or partly known to some students. The course was specific by its content, in comparison to the content of some other IT course (for example in the field of programming), this content was not too complex for independent learning and was also suitable for the attractive formation into the web materials for online learning.

Given that some topics were known to students, they could decide for themselves which lessons they would read and study in depth, and which they would not. In this way we avoided the situation that often occurs in the f2f teaching where all students are obligated to listen to all the topics presented in the course, even though the topics might be familiar. On the other hand, students that did not know some introductional topics that they should have already known, had an opportunity to repeat those topics (for example, part of the topic about the WWW). In order to individualize the course and encourage the students to develop an active approach in the knowledge acquisition, the technology was irreplaceable.

2.2 Learning Objectives and Content

The overall objective was that students acquire fundamental knowledge about the digitalization of a single media (graphics, text, sound, animation, and video) and integration of these media into a multimedia project.

On completion of this course, students should be able to:

- define and compare the concepts of multimedia, hypermedia and hypertext,
- outline and explain advantages and disadvantages of multimedia and hypermedia,
- outline, describe and compare digital media elements: graphics, text, sound, animation, and video,
- develop and design simple digital multimedia files: graphics, sound, animation, and video clips,
- organize multimedia elements into web presentation by WWW standards and according to the phases for multimedia project development.

The course consisted of the following theoretical topics: Introduction to multimedia, hypertext, and hypermedia, Main concepts of the World Wide Web, Graphics, Sound, Video, Animation, Text, Multimedia projects development.

The topics for the exercises were connected with the lectures: Creation of a web site, HTML, CSS, XML, SVG, Image processing (Photoshop), Sound processing (Audacity), Making and processing of a video (Windows Movie Maker, Screencast-O-Matic), Making animations (Flash), Combining multimedia elements in a Flash presentation.

2.3 Approach to Blended Learning and Learning Activities

2.3.1 Sequential Blended Learning Model

Blended learning (BL) is becoming an increasingly popular form of e-learning, particularly suitable for use in the process of transition from traditional forms of learning and teaching towards e-learning [5, 7]. In this model of teaching and learning, significant amounts of f2f elements are replaced by technology-mediated teaching. Therefore, fewer f2f class sessions are held because ICT is increasingly being used to deliver course materials and to facilitate learning. According to [5], the most efficient teaching model is a blended approach, which combines self-paced learning, live e-learning, and f2f classroom learning.

This model was applied to the course "Multimedia systems", and was based on a combination of independent learning, f2f practical work on computers, online discussions, and problem-based learning (making a seminar – multimedia presentations). The model uses both f2f environment and an online environment supported by a learning management system (LMS). In this course, an active approach in teaching and learning was encouraged, with the instructors not „delivering" knowledge but directing and encouraging students to acquire it actively.

F2f lectures were not obligatory for students. Instead, the instructor offered an opportunity to the students to listen to the summary of the topic in a form of a PowerPoint presentation in which the key concepts of the topic were pointed out. In case only a few students came to this optional class, consultations took place.

What was obligatory was to attend the exercises for which students had to prepare themselves in advance according to the published materials and solve the tasks during the class with an individual help of the assistant offered to the students that needed it.

The course was prepared according to the sequential (program flow) model of blended learning [2], mainly because this model was particularly appropriate in transition from standard f2f teaching to a model which introduced online learning [7]. It is important to emphasise that not only the online and classic f2f environment are connected, but also the media for delivering the content for learning as well as different teaching and learning methods that are integrated chronologically in the course. The course is consisted of steps that are executed in the exact order and through which all students go through lineary. The course schedule is known from the beginning until the end of the course which is published in the course calendar. At the end of some cycles and the whole course, a summative assessment takes place. The order of the activities is shown in the Figure 1.

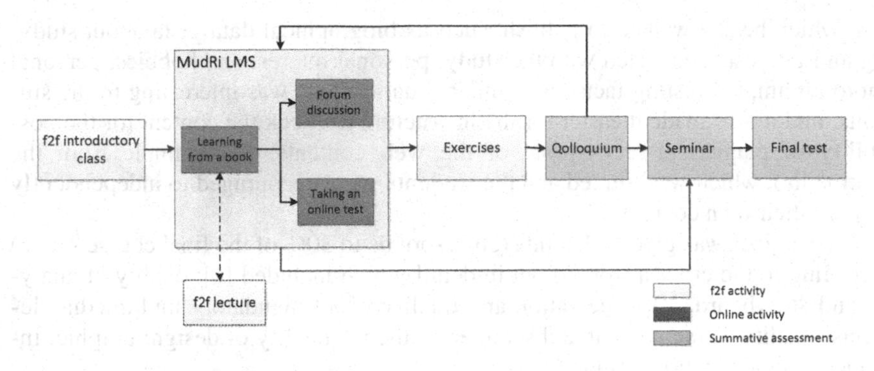

Fig. 1 Activities followed during the course.

The course commenced with a two-hour kickoff f2f lecture where the learners had the opportunity to meet each other and the teachers. This f2f introductory session was dedicated to the presentation of the learning objectives, discussion of the most significant knowledge and tasks to be learned, and description of all learning activities. At the beginning of the course students were given the course password by which they could enrol into the Moodle course.

The Moodle course was consisted of the set of blocks that were not published at once, and were further developed depending on the predefined schedule of the class. Activities were announced in a form of a forum message every week.

Every week or every other week (for more extensive topics), a corresponding block for theoretical topics with the appurtenant learning outcomes, a Moodle book, online tests for self-assessment, and a summary in a form of a PowerPoint presentation was opened. The instructor also published a group of questions in a Question & Answer (Q&A) forum.

For the exercises, the materials were published in the block „Exercises", also during the week for the corresponding lecture. The content of the exercises was in compliance with the theoretical topic in the corresponding week so that all learning outcomes could be completely achieved.

It was expected from the students to first study the theoretical content, which most important elements could be strengthen with the help of the questions published in the forum. Students could check their knowledge in an online test for the self-assessment. In the continuation of the learning, students had to study the materials for the exercises in order to successfully solve the given tasks during the class.

2.3.2 Seminar

Students had a task to make an extensive seminar – a Flash presentation. This is a multimedia presentation made according to the given topic in the program Flash by using the knowledge gained through the semester. Students were given further instructions about the topic, the seminar and the way of grading the seminar. In the academic year 2010/2011 the topic was a personal biographical presentation of a student (a form of a multimedia ePortfolio) for which a student had to present the data which he/she wanted to publish, such as biographical data, data about studying and interests connected with the study, personal interest and hobbies, personal photo album, interesting facts and similar. Such a topic was interesting to the students, and it also made it easier for the instructors to check the content for the possibility of plagiarism (copy-paste of the web contents, for example from the Wikipedia), which was limited and the students were encouraged to independently prepare their own content.

The seminar was given 30 points (corresponds to 30% of the final course grade) according to the criteria worked out in detail which included [6]: quality of analysis and storyboard, implementation and quality of all mandatory multimedia elements, quality of the content and seminar elements, quality of design: graphic, interface, and navigation design.

2.4 Knowledge Assessment

Summative assessment was held after half of the topics were covered in the class in a form of a colloquium (30 points) and at the end of the course in a form of a final test (30 points). In both cases, students had to solve online test and several practical assessments. They uploaded solutions of practical tasks in the LMS so that the instructor could assess them. In the same way, students submitted their seminars.

Based on a results given by checking student's activity logs, reading the content in the system MudRi, solving online tests for the self-assessment, and participating in course forum discussions, students could get up to 10 points. In this way students were encouraged to continually and independently work in the LMS system, and especially communicate on the forum.

The final grade was given on the basis of summation of all gathered points during the course by participating continually in the given activities and taking the final test according to the following scale: A - 80-100% - excellent (5), B - 70-79.9% - very good (4), C - 60-69.9% - good (3), D - 50-59.9% - satisfactory (2).

3 Used Technology

The e-course was implemented in the Moodle LMS MudRi system for learning management University of Rijeka. The elements of the MudRi system that were used in the course are: books, tests for self-assessment and knowledge assessment, home assignments that use advanced file submission, forums (for announcements, discussions and Q&A forums), a calendar, grades report, survey, reports. The course used a thematic form which consisted of the following blocks/topics:

- announcements and discussions – contained a forum with announcements, a forum for student's questions and discussions and special Q&A forum for students
- information about the course – contained a detailed curriculum, as PDF presentation from the introductory class and important information (such as test dates)

- a survey – a survey that was completed at the end of the course
- up to one block for every course topic: Introductory concepts of multimedia and hypermedia, World Wide Web, Graphics, Text, Audio, Video, Animation, Development of the multimedia projects
- exercises – a block that contained the materials for preparation, examples and tasks for independent solving during the class or as a home assignment
- a seminar – contained instructions for making a multimedia presentation in a program Flash, grading criteria, and the possibility to submit certain parts (storyboard) of the seminar and the finished Flash presentation
- blocks: Colloquium, Final test, Makeup exam – online tests for assessment of theoretical knowledge with instructions, tasks for practical part with corresponding files and submission of the solution.

The main elements for the online part of the course were blocks used for the course topics. Every block contained: elaborated learning outcomes for the topic, a list of the literature and links used for the preparation of the content (they also served the students who are interested into an additional source of literature), a Moodle book with lectures connected with the topic, a test for self-assessment, a PDF document with a PPT presentation given at a f2f lecture.

A forum with announcements serves the instructor to announce a corresponding topic and all other announcements and important information every week. As the forum allowed only instructors to post messages, a new forum *Questions and discussions* was created, in which students could comment and respond to the instructors messages, publish messages based on different topics connected with the course content and encourage the discussions among themselves.

Q&A forum was specific for the fact that a student had to answer the proposed questions given by the instructor in order to see other student's answers. Considering the fact that there were a lot of students, everyone got a chance to show their understanding of the content and write their own answer regardless of when they began participating in the discussion. After the student answered the question, he could continue commenting on other student's answers.

In making the tests, we used all Moodle types of questions, although most of them were multiple choice with one or multiple correct answers followed by a gap fill type of questions. In test for summative assessment, we used the essay-based questions in which students had to write in a more detailed answer which were not automatically graded but evaluated by the instructor. There were about 300 questions created for the course, organised into 60 categories which corresponded to each lecture written in a Moodle books.

The tool for advanced submission of assignments allowed students to hand in their files connected with the seminar or a colloquium and a test. Apart from the detailed learning outcome plan, schedule and announcements in the forum, the flow of the course activities was also published in a calendar.

Anonymous survey allowed students to express their thoughts about the course and the blended e-learning model (online and f2f) with the help of the MudRi system.

With the help of the grades reports, every student and the instructors could follow the points for each activity during the course, final points and a final grade. The activity reports helped the instructor to keep track of the student's progress in MudRi during the course and give 10 points for this element of student evaluation.

3.1 Efficiency of the Used Technology to Achieve Student's Motivation

Characteristically for the students enrolled in the course was that these were informatics students that are used to digital media and the MudRi or some other LMS system which they had been using for three years, since they started with their studies. In this sense, the used technology was not by its self new, but it was important to design activities that would encourage students to learn instead of using the MudRi as a repository of the digital content (such a case can be found in majority of other courses in which these students are enrolled).

Majority of informatics students (both undergraduate and graduate studies in which the instructor holds a couple of courses in which the same student group is enrolled) had good skills in working with the technology (program tools), but when it comes to being familiar with the theory, students "practitioners" were not used to learning from the written materials (classic, printed or online books). A special problem is students' incapability to concisely and clearly express themselves in a written form. Therefore, answering the questions posted in a forum was an endeavour to encourage the students to develop communication skills in a written form which belong to basic competencies of a future IT specialist.

Based on the observation during the course and the results of the student survey, the forum was proven to be the less desired activity. Apart from an extrinsic motivation (points given for the participation in the activities), we tried to encourage students to participate in such communication by constant encouragement.

It is also important to mention that this way of working in the course demands a continuous student work through the semester, but due to other obligations and different circumstances not all students obeyed the schedule. Therefore, some activities (such as answering the questions posted in a forum and solving the online tests) were more intensive prior to the colloquium and at the end of the semester when students learned couple of topics at once.

4 Evaluation

In the academic year 2010/2011, 29 students were enrolled in the course "Multimedia systems". Depending on their academic results, the course pass rate was 69% with an average grade 3,31 (62% points). Almost all students passed the course test in the first, summer term.

The evaluation in a form of an anonymous Moodle survey was made at the end of the course with a goal to establish in which measure students accepted this model of learning (based on the BL paradigm) and especially which activities and elements of the e-course they considered the most useful for learning. The survey was prepared based on the years of instructor's experience in questioning a student's attitude on online learning or blended learning [6].

In one part of the survey, the students were asked to express their opinion using the 1–5 Likert scale (1= strongly disagree, 5= strongly agree with the statement). In other questions students chose the activity or tools for which they considered to be of most use.

The anonymous survey was completed by 20 students (69% students who were enrolled in the course). The results were not statistically significant, but helped the teachers in deciding where to focus future efforts and how to continue with the development of blended learning model for the course.

4.1 Questionnaire Results and Student Comments

Based on the survey results, it can be said that students accepted the blended learning in a good way. The most interesting statements from the survey are presented in Table 1.

Table 1 Survey results

	Avg.	StDev.
You are satisfied with the new way of working during the course which uses e-learning.	4,15	0,67
You consider the new way of working efficient for learning.	3,95	0,83
In comparison with the traditional teaching, You needed more time to study while using the MudRi system.	2,3	0,98
In comparison with the traditional teaching, the new way of e-learning demands more personal engagement from Your side.	4	0,92
We would gladly use e-learning and MudRi in another course.	3,95	0,89
The learning materials for the course were written clearly and legible, with nice colours and were formatted in a simple and standard way.	4,05	0,94
The learning materials and activities on the lectures and exercises were organised good.	4	0,97
The availability of the obligatory literature in a form of a book with chapters helped me to master the course materials.	3,8	1,20
Self-assessments helped me to master the course materials.	4,2	0,70
Working on the seminar helped me to master the course materials.	4,15	0,88
Online forum discussions helped me to master the course materials.	2,74	0,93
The instructor regularly updated the contents and managed the activities of the course in the MudRi system.	4,5	0,76
The instructor communicated with the students through the MudRi system often enough.	4,4	0,82

The results of the survey show that the forum was the only activity which was less accepted from some students. The survey also revealed that all other elements were useful to the students, for example 75% of the students chose self-assessment as the most useful elements of the course, 55% seminar, 60% books with course content. Although no one singled out the forum as the most useful element, 20% chose all listed elements in a whole (including a forum) and the same amount agreed that the forum discussion helped them to master the course materials. Still, students considered that they learned the most by working on the seminar (60%), solving the quizzes (55%) and learning from the books (50%).

Special value to the survey was given by students' comments on what they most liked in the course and what they thought should be changed. Some of the mentioned advantages were:

- The course was very interesting and fun, which affects the overall impression of e-learning.
- We could organise the time and a place of learning on our own.
- The concept of the independent learning without a physical presence in the classroom and a possibility to chose the content which we want to learn.
- Constant availability of all materials.
- Tests for self-assessment because they offered us to test our true knowledge.
- Communication between a professor and a student through MudRi forum.
- All information needed to successfully pass the course was available. "I liked this way of "forcing" the students to constantly work through the semester, and not just before the colloquium."
- Students are more self-independent and they organise their work on their own. The theory is not too difficult and it is clearly written in books. The instructor was available for consultations and questions if something was unclear.

When asked if there was something they would change, some students stated that this way of learning required too much time and that the instructor required excessive participation on the forum, which, in their opinion, should not be mandatory. Some further proposals for the course improvement referred to adding more tests for self-assessment and a possibility to print out the materials in a PDF format.

The need for the model improvement arises from the fact that students do not prefer to communicate in discussion forum. In current situation at the University where most of the instructors use f2f lectures, and a small number of them use LMS, students are not used to written communication. In addition, implemented version of the Moodle LMS does not include Web 2.0 tools that would be potentially more attractive to the students and enable collaborative learning. Therefore, in our plan is to combine the LMS with 2.0 tools available on the web. By introducing Web 2.0 tools such as wiki, blog, and ePortfolio we will tend to foster activities like writing individual diary and collaborative writing, and try to increase students' motivation for participation in such activities.

5 Conclusions

In this paper, a sequential model of the blended e-learning for the course "Multimedia systems" which was implemented by the combination of the online and f2f classes with an emphasis on self-paced learning, online discussions and making the seminar was described. Educational activities, use of technology, and final results are illustrated.

Based on the evaluation survey it can be concluded that the introduced model for blended learning was successful and that students were mostly satisfied with this way of working. We will single out one student's comment, who stated that he "thinks that it depends only on students how much this way of working will be used." This sentence expresses also the attitude of the instructor on managing a course in a new way. In a blended model which was used in the course, students are offered different activities and tools, both f2f and online, so that every student can find more elements that are fit to him and achieve the learning outcomes for this course.

It is in our plan to further develop and improve the described blended learning model and extend online learning environment with Web 2.0 applications. Special attention will be given to further encourage activities which will help students to develop literacy in native language, which was partly neglected while educating the future computer scientist.

References

1. Babb, S., Stewart, C., Johnson, R.: Constructing Communication in Blended Learning Environments: Students ' Perceptions of Good Practice in Hybrid Courses. Learning 6(4), 735–753 (2010)
2. Bersin, J.: The Blended Learning Handbook. Wiley, New York (2004)
3. Commission for e-learning, University of Rijeka, Strategies of implementation of e-learning at the University of Rijeka 2006-2010 (2006), http://www.uniri.hr/files/staticki_dio/propisi_i_dokumenti/Strategija_uvodjenja_e-ucenja_UNIRI.pdf (accessed January 22, 2012)
4. European Association for Quality Assurance in Higher Education (2005) Standards and Guidelines for Quality Assurance in the European Higher Education Area (ESG), http://www.ond.vlaanderen.be/hogeronderwijs/bologna/documents/Standards-and-Guidelines-for-QA.pdf (accessed January 19, 2012)
5. Garrison, D.R., Vaughan, N.D.: Blended learning in higher education: Framework, principles, and guidelines. Jossey-Bass (2008)
6. Hoic-Bozic, N., Mornar, V., Boticki, I.: A Blended Learning Approach to Course Design and Implementation. IEEE Trans. Edu. 52, 19–30 (2009)
7. Napier, N.P., Dekhane, S., Smith, S.: Transitioning to blended learning: Understanding student and faculty perceptions. Journal of Asynchronous Learning Networks 15(1), 20–32 (2011)
8. Precel, K., Eshet-Alkalai, Y., Alberton, Y.: Pedagogical and Design Aspects of a Blended Learning Course. International Review of Research in Open and Distance Learning 10(2), 1–16 (2009)
9. The European Higher Education Area, Joint Declaration of the European Ministers of Education (1999), http://www.eua.be/eua/jsp/en/upload/OFFDOC_BP_bologna_declaration.1068714825768.pdf (accessed January 19, 2012)

Prototyping an Online Game Platform through the Formative Design Approach Based on the Monopoly Mechanism

Yu-Hui Tao[1], Wei-Jyun Hong[1], and C. Rosa Yeh[2]

[1] Department of Information Management, National University of Kaohsiung
[2] Graduate Institute of International Human Resources Development, National Taiwan Normal University

Abstract. The goal of this research is to implement a prototype online game learning management platform with high usability via a formative design approach. Two stages of usability evaluations, namely, cognitive walkthrough and heuristic evaluation, are applied to assess the prototype system, followed by a small-scale survey. A total of 15 suggestions are obtained; In general, the students adequately perceived the incremental functionalities on real estate transactions and the learning concepts intended by this research. As a learning game, its design earned recognitions and raised user intention to use this platform as a learning tool. The overall user satisfaction on both the game and the class management platform ranges from neutral to slightly positive, thereby implying a positive general user attitude with more expectations for improvements. With this initial formative evaluation, the outcomes provide valuable input to subsequent formative design stages, as well as a foundation for high usability, closer to actual game scenarios, and a pioneering platform for informal game learning in higher education applications.

Keyword: business simulation game, formative evaluation, learning platform, Monopoly.

1 Introduction

Several studies have repeatedly shown that digital learning games can achieve better teaching outcomes (Coller & Scott, 2009). However, although many business simulation games meet the learning objective, some often ignore the fun of game playing (Tao et al., 2009). Therefore, game designers should find the right balance between playfulness and the learning objective. Meanwhile, perceived ease of use is a critical factor for system development, which is particularly important for students' continual usage of the business simulation games (Tao et al., 2009). In addition to the game itself, majority of business simulation games, such as the most popular Business Operations Simulation System (BOSS) in Taiwan, focus on game playing and competition, but lack in class management functionality, such as the e-learning platform commonly used in colleges.

L. Uden et al. (Eds.): Workshop on LTEC 2012, AISC 173, pp. 77–88.
springerlink.com

Monopoly is a world famous strategic game for all ages. In fact, according to Axelrod (2004), this game initiated his knowledge on real estate transactions. Accordingly, such scholars as Yeh et al. (2007) applied paper-based Monopoly game activities in the class "Dynamic Strategic Planning," and Tao et al. (2010) proposed an improved Monopoly online game design as a business simulation game with integrated management platform functionality. Therefore, Monopoly is an ideal solution to address the first issue of enhancing the playfulness of traditional business simulation games as well as the third issue of improving class management functionality. To implement an easy-to-use Monopoly game, as described by Tao et al. (2010), we adopted a formative design and evaluation approach to address the second issue of improving usability in this research.

2 Brief Background

Strategic games can be divided into real time strategy and turn-based strategy (King & Krzywinska, 2002). Monopoly is a multi-person turn-based strategic game. In Taiwan, college students have adopted Monopoly concepts in their campus activities. Examples of games patterned after Monopoly are Feng-Chia Monopoly (2009) and Tsing-Hua Monopoly (2009). Feng-Chia Monopoly, which was designed by graduating student committee from Feng-Chia University, included campus scenes for graduating students. Meanwhile, Tsing-Hua Monopoly, which was designed by Electronic Engineering students from Tsing-Hua University, featured campus scenes and important characters such as past Presidents.

The official developer of Monopoly is Parker Brothers, which has continuously designed new and different versions of the original game. Tao et al. (2010) proposed a version, which aims to develop the game into one that is closer to the actual environment so that students can learn domain knowledge, such as real estate transactions featured in the original game. In addition, this new Monopoly-patterned game is integrated within the framework of a class management system, such that playing the game as assignments can be managed by a platform similar to the e-Learning platform used by teachers in the same institute.

3 Research Design

The research design includes the implementation and formative designs, followed by the evaluation. The design of teaching management does not refer to e-Learning platforms, such as Moodle, but to a specialized teaching platform called Peer Evaluation System (Tao et al, 2007), which has been developed to manage assignment evaluations by multiple teachers and multiple classes. In addition to the common functions (e.g., management of account, assignment, and evaluation), specialized functions for game playing (e.g., management of game design, team, and game playing) are added. The process by which teachers and students proceed with the class assignment activities through this class management subsystem is illustrated by the use case diagram in Figure 1.

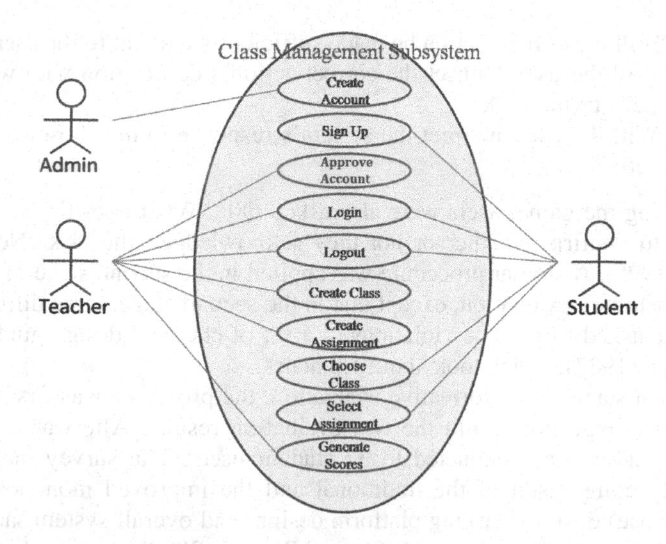

Fig. 1 Use case diagram for class management

The Monopoly game design used the conceptual design by Tao et al. (2010), with the experiences from the original version by Parker Brothers (http://www.hasbro.com/monopoly/en_US) and Softstar in Taiwan (http://richol.joypark.com.tw). To develop a closer to actual real estate environment, additional features were introduced: (a) price index function, so that the costs and fees can fluctuate during the game; (b) more complete transaction types by including mortgages and auctions; and (c) flexibility of the number of houses that can be build, in accordance with the owner's intention, instead of fixing the number to one house at a time. For the prototype to be easily accessed from web browsers, Java Servlet, JSP, and Java Script languages were used, together with other freeware, such as MySQL 5.5 database, Tomcat6.0 web server, Eclipse-integrated development environment, and JQuery and Ajax packages. In particular, Direct Web Remoting (DWR) 2.0 was used to enable the server push mechanism for players to simultaneously see the same screen changes, which was not possible in traditional HTML environment. As an initial stage of a formative design process, the prototyping and evaluation was divided into three stages: (i) cognitive walkthrough method for functionality improvement; (ii) heuristic evaluation method for performance improvements through design guidelines; and (iii) student perception survey after experiencing the prototype for a summative evaluation.

Once the prototype reached a certain level of completion, the cognitive walkthrough was conducted, and three graduate students enrolled in the Man-machine Interaction course were invited as evaluators. The experiment took about 1 hour, including a 10-minute warm up, 30-minute game play for 50 rounds, and the remaining time for cognitive evaluation assessment. A total of 10 game screens were captured in asking the following questions (Newman & Lamming, 1995) to each evaluator:

(Q1) "Will the correct action be made sufficiently evident to the user?"
(Q2) "Will the user connect the correct action's description with what he or she is trying to do?"
(Q3) "Will the user interpret the system's response to the chosen action correctly?"

Before playing the game, users were also asked Q0 ("What does the user want to achieve?") to confirm whether or not they acknowledged the task (Newman & Lamming, 1995). A similar procedure was applied in the second stage of the experiment for heuristic evaluation, except that in the second stage, three different evaluators were asked to provide violations of a set of classical design guidelines by Shneiderman (1987), which took about 1.5 hours.

In the final stage of the formative evaluation, the prototype was first improved based on the suggestions from the two evaluation results. Afterwards, an after-experience survey was conducted to 32 student users. The survey included the functionality comparison of the traditional and the improved monopoly games, game (interface) design, learning platform design, and overall system satisfaction. Studies by Olivas and Newstorm (1981) and Ranalli (2008) were used as bases in designing the survey items. The items included multiple choices, fill-in-the-blanks, and a nine-point Likert scale, which is the most widely used psychometric scale in survey research (http://en.wikipedia.org/wiki/Likert_scale).

4 Formative Evaluation

4.1 First Phase: Cognitive Walkthrough Evaluation

In this experiment, 10 game screens were captured so that the three evaluators can discuss the design flaws related to Q1 to Q3, with possible suggestions for improvement. Symbols A, B, and C were used to represent the three evaluators, respectively. An auction bidding screen was used to illustrate the results before summarizing all the problems raised by the three evaluators.

Figure 2 captures the choice of one player for auctioning the land, where the other players can input the bidding amount in the popup screen. The popup window wrapped in the red rectangle includes the land auction information, three bidding prices buttons (i.e., +20, +50, and +100), and the time counter on the bottom right corner which shows the remaining time (5 seconds). The players can input their bidding amount freely before the time runs out; the player with the highest bidding amount wins the land auction.

In this screen, only evaluator A has raised issues for all three questions, which implies extreme perceptions among the evaluators. Evaluator A raised several issues: (i) for Q1, the wording presentation on the button needs improvement to make it more intuitive to the users; (ii) for Q2, the rule for winning the bid is not clear; and (iii) for Q3, the feedback showing the winner is not clear since the window closes after the bidding time (Q3).

The corresponding suggestions for changes include the addition of dollar sign ($) in the bidding text on the buttons and extension of the life cycle of the popup auction window after the bidding time, so that all playes can have sufficient time to know the bidding results.

4.2 Second Phase: Heuristic Evaluation

Based on the eight general design guidelines by Shneiderman (1987), three evaluators (coded as D, E and F, respectively) discussed their perceived violations of the guidelines individually based on their prototype game playing, after which they made suggestions for improvement. The design guidelines include "strive for consistency," "enable frequent users to use shortcuts," "offer informative feedback," "design dialog to yield closure," "offer simple error handling," "permit easy reversal of actions," "support intentional locus of control," and "reduce short-term memory load." The most discussed design guideline, "reduce short-term memory load," was used to illustrate the results before summarizing all the problems raised by the three evaluators.

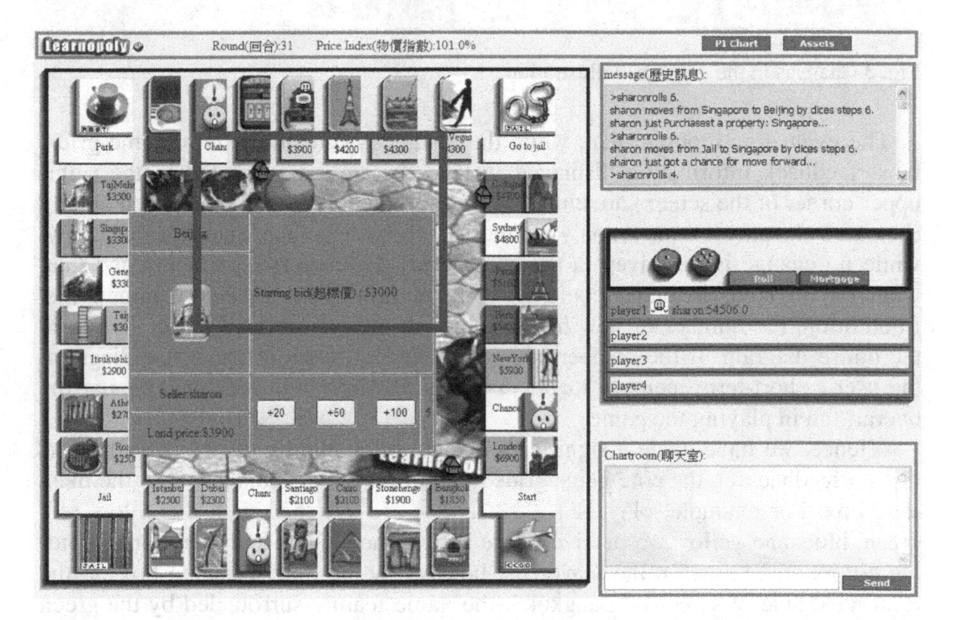

Fig. 2 Auction bidding screen

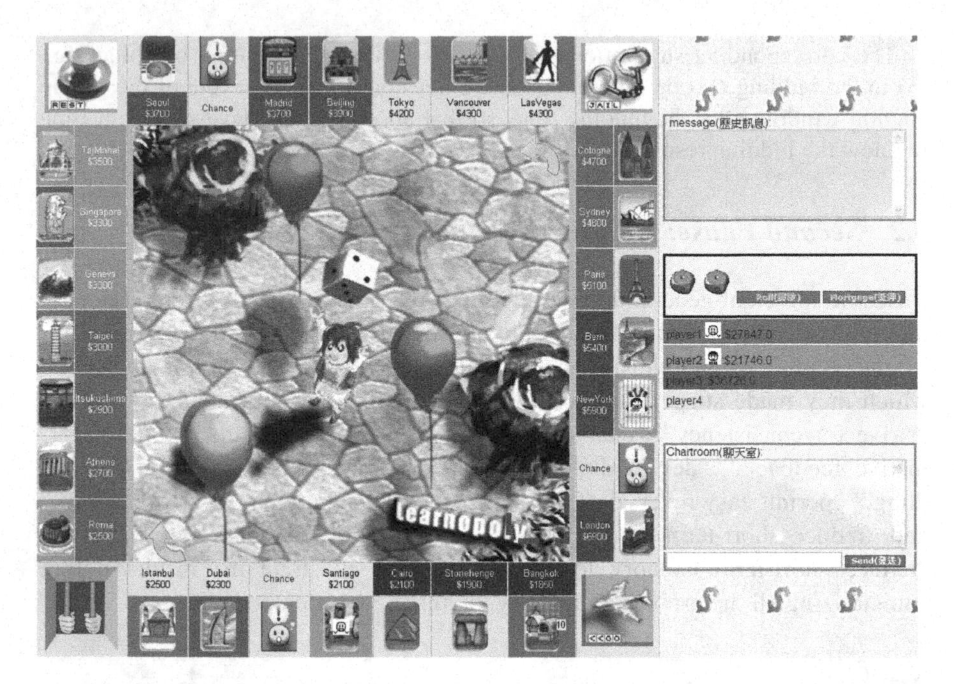

Fig. 3 Changes in the game grid illustration

The evaluators reported that, while the changes occur in the core game grids, most feedback information displayed in the message box (located in the right-upper corner of the screen) are either flashed too quickly or are not labeled clearly (i.e., after the player's move, only the message box shows the change accordingly, while no change is perceived in the game grid). Evaluators also mentioned that, when buying land, the player could not confirm directly from the core game grids. In addition, the number of new houses built on the private land is not shown in the house diagram. If the number of houses can be shown in the house diagram, the user's short-term memory load can be significantly reduced, thus raising the overall fun in playing the game.

Hence, we have made a significant change by explicitly displaying all information feedback on the core game grids in addition to those displayed in the message box. For example, players 1, 2, 3, and 4 are assigned with the colors red, green, blue, and yellow, respectively (see Fig. 3; the players and their corresponding colors are shown on the player list box below the dice throwing box on the right hand side of screen). "Bangkok" (the game feature surrounded by the green dotted box in the bottom right corner), has a red background color, indicating that it is owned by player 1. Furthermore, there are 10 houses built for Bangkok by player 1, as indicated by the number 10 appearing in the grid next to the house diagram.

Table 1 Issues identified and the corresponding status

Evaluation Phase	Identified issues
First	Action feedback is not clear
	Zero should be the default value in the input field for buying a house
	The label text on the "submit" button is not sufficiently intuitive
	Lack of explanation for mortgage
	The label text on the "bidding" button is not sufficiently intuitive
	Lack of auction rule explanation
	Information regarding the bidding winner after auction is not clear because the window closes immediately after the auction
	Bankrupt causes game halt
Second	Lack of feedback after action or undo functions
	No shortcuts for game playing
	More intuitive information feedback should be provided
	No error handling*
	Lack of function information on button labels
	Lack of intuitive information feedback on land buying
	Lack of intuitive information feedback on house building

* unfixed

The major issues identified by the six evaluators in both cognitive walkthrough and heuristic evaluations are summarized in Table 1, with confirmations of corresponding changes. The changes are based on the solutions suggested during the discussion with the six evaluators, when the issues have been raised. Therefore, most changes are thoroughly discussed by the corresponding three evaluators with acceptable levels of usability.

4.3 Third Phase: Small-Scale Summative Evaluation

A total of eight game groups, each with four players, were employed in this summative evaluation. To assess the internet game-playing and learning management functions, six games were tested via the Internet, in which the players continued playing the game on their own computers at home after completing the first two games in the laboratory. In our sample of 32 subjects, the percentage of male participants (81.25%) is considerably higher than that of the females. In terms of age, majority of are between 20 to 30 years old. All participants attained college and postgraduate degrees, all of them have game playing experiences, but only four have digital learning game experiences.

Table 2 Comparison of the traditional and improved versions

Monopoly Components	Traditional (1)		This research (2)		% Difference (2) to (1)
Event	Sample size	%	Sample size	%	
Land purchase, rent, jail, chance, community chest	160/160[a]	100	157/160	98.13	-1.87
Number of houses	2/32	6.25	27/32	84.38	78.13
Price index	5/32	15.63	30/32	93.75	78.12
Auction	10/32	31.25	27/32	84.38	53.13
Mortgage	12/32	37.5	15/32	46.88	9.38
Property sale	36/96	37.5	55/96	57.29	19.79

[a] 32 data points/event x 5 events=160 data points

The first part of the questionnaire compares the perceived functional differences between the traditional and improved Monopoly games. As presented in Table 2, most frequently seen Monopoly components, namely, "land purchase," "rent," "jail," "chance" and "community chest," are perceived as nearly equivalent on both versions. The new functions introduced in our improved game obtained mixed perceptions. More than 85% of the users recognized the fact that flexible number of houses and a priced index are made available in our improved version. Three perception differences are considered significant. First, only 30% to 40% of the subjects understood the mortgage function in both versions, and 10% more had the same reaction to the improved monopoly game. Second, nearly 85% of the users perceived the "auction" function in the improved version, whereas only 31% perceived it in the traditional version. Finally, 37% perceived "property sale" in the traditional version, whereas 57% perceived it in the improved version.

For the first set of components (i.e., land purchase, rent, jail, chance, and community chest), all other items are perceived higher in our improved version than in the traditional game. Thus, the improvements in our version elicited significant feedback from the users.

User perceptions regarding the real estate transaction, game design, and overall satisfaction were also investigated. The average perceived scores ranged between 5 and 6, indicating neutral to slightly positive levels. Although this result is less ideal, it implies a positive feedback toward this initial formative design and evaluation process, establishing a good foundation for the consequent formative design process. The three remaining parts of the questionnaire are discussed separately in the following part of this section.

For the game comparison, the construct average for the concept of real estate transaction is 6.01, which is an acceptable positive level (see Table 3). The users perceived our version of the game as being closer to the real world than the traditional version (Q1); however, it is still slightly different from the actual real estate environment (Q2). The average score for Q3 (6.28) implies that our version clearly expresses the concepts of actual real-estate transactions.

Table 3 Concept of real estate transaction

Item #	Description	Average	Construct average
Q1	The new Monopoly learning game is closer to actual real estate transactions than the traditional version.	6.25	
Q2	The new Monopoly learning game is very close to actual real estate transactions.	5.5	6.01
Q3	The new Monopoly learning game efficiently expresses the concepts of real estate transactions.	6.28	

Table 4 shows the results for the game design. The construct average is only 5.36, indicating that much improvement can still be made. The highest average of Q4 indicates the choice of using Monopoly is a good one since almost all of the participants know how to play the game. Meanwhile, the lowest scores are obtained for Q7 and Q9, indicating necessary improvements regarding better graphic designs and visual effects as well as the use of a robust game software, respectively. The latter is attributable to the constraints of older freeware version; for instance, DWR 2.0 can only be used with IE 7 browser.

Table 4 Game design

Item #	Description	Average	Construct average
Q4	The rules for playing the new Monopoly learning game are very clear.	5.97	
Q5	The new Monopoly learning game provides the players clear information while playing the game.	5.47	
Q6	The new Monopoly learning game provides the players adequate flexibility for sections.	5.69	
Q7	The visual effects and graphic designs of the new Monopoly learning game are commendable based on the 2D perspective.	4.88	5.36
Q8	The new Monopoly learning game provides the player good manipulation capabilities.	5.56	
Q9	The new Monopoly learning game is very robust, and no problem occurs easily.	4.59	

Table 5 presents the results for the overall game satisfaction. The average construct is 5.53, which indicates that the features should still be further improved. Among the six items, similar to Q4 (see Table 4), Q10 obtained the second highest score, indicating that the popularity of the game makes it easy to play. Items Q12 and Q13 are conceptually related; their results suggest that the new version, although not as fun as its commercial online counterparts, is entertaining enough to attract students as a learning game. Items Q14 and Q15 are also related, with results indicating that users may not want to continue using it for learning real estate transactions but would use it for teaching others instead. These two relative comparison item sets imply that online learning games, such as our Monopoly game, are still needed.

Table 5 Game satisfaction

Item #	Description	Average	Construct average
Q10	The new Monopoly learning game is easy to play.	5.81	
Q11	The new Monopoly learning game assists students in learning many concepts and terminologies involved in real estate transactions.	5.69	
Q12	The new Monopoly learning game is as fun as the commercial online games.	4.94	
Q13	The new Monopoly learning game is a fun game to play for learning purposes.	5.88	5.53
Q14	In the future, if I want to review knowledge of real estate transactions, I will consider playing this game again.	5.28	
Q15	In the future, if I want to teach knowledge of real estate transactions, I will consider using this game.	5.56	

Table 6 shows the results for the platform design and satisfaction. The construct average is 5.54, which is similar to result for game satisfaction, as discussed above. All items scored higher than the neutral value of 5, implying a positive attitude perceived by the users; however the game can still be further developed. Among the four items, Q17 attained the lowest average, which reflects that more functionality is required to make our Monopoly version closer to what the participants have experienced on the e-Learning platform. Items Q18 and Q19 obtained scores that have the same implications as Q14 and Q15, that is, this management platform is still preferred for online learning games.

Table 6 Platform design and satisfaction

Item #	Description	Average	Construct average
Q16	The new Monopoly learning game platform is easy to comprehend.	5.66	
Q17	The new Monopoly learning game platform provides comprehensive functionality.	5.38	5.54
Q18	The new Monopoly learning game platform is appropriate for assisting student learning activities.	5.59	
Q19	In the future, I would like to use this Monopoly learning game platform.	5.53	

5 Conclusions and Future Work

This research contributes to digital learning games in two perspectives. First, we implemented an online learning game platform by integrating a class management system, which is not seen in the learning game environment, into an improved Monopoly game which is closer to actual real estate transactions for learning purpose. Second, we adopted a formative design and evaluation approach at the initial stage of this prototype system for increased usability.

We conclude that an online learning game with an integrated class management system is technically feasible and is preferred by the users. In addition, the formative design and evaluation process assure adequate usability at the initial stage of the system development, with at least neutral to slightly positive user perceptions. This prototype system can be used in management-related courses for student majoring in management or all students participating in general courses.

These sustained research outcomes provide a solid foundation of user acceptance to subsequent formative designs and evaluations for future improvements. For example, to improve user satisfaction, given that six groups of participants reported complaints regarding inconvenience related to freeware constraints, newer versions of development tools should be employed. The ultimate goal of this Monopoly-patterned online learning game platform is to realize the learning-by-design concept, which is a more advanced game learning-by-doing concept, as proclaimed by Prensky (2008) and Lim (2008). Therefore, to meet the objective of "learning by design," a Monopoly game design engine is desirable, because students can easily design a customized Monopoly-patterned game as an assignment.

Acknowledgments. This research was sponsored by National Science Council of the Republic of China under Grant No. NSC 97-2410-H390-012-MY3.

References

Coller, B.D., Scott, M.J.: Effectiveness of using a video game to teach a course in mechanical engineering. Computers & Education 53(3), 900–912 (2009)

Feng-Chia Monopoly, Public Relation News (June 9, 2009), `http://sdsweb.oit.fcu.edu.tw/FcuNews/contents.jsp?id=B94CED750D3397E3482575D0004E945F&lang=C` (accessed February 12, 2012)

King, G., Krzywinska, T. (eds.): ScreenPlay: Cinema/Videogames/Interfacings. Wallflower Press, London (2002)

Lim, C.P.: Spirit of the game: Empowering students as designers in schools. British Journal of Educational Technology 39(6), 996–1003 (2008)

Newman, W., Lamming, M.: Interactive System Design. Addison-Wesley, Reading (1995)

Olivas, L., Newstorm, J.W.: Learning through the use of simulation games. Training and Development Journal 35(9), 63–66 (1981)

Prensky, M.: Students as designers and creators of educational computer games: Who else? British Journal of Educational Technology 39(6), 1004–1019 (2008)

Ranalli, J.: Learning English with the sims: exploiting authentic computer simulation games for L2 learning. Computer Assisted Language Learning 21(5), 441–455 (2008)

Shneiderman, B.: Designing the user interface: Strategies for effective human computer interaction. Addison-Wesley, Reading (1987)

Tao, Y.-H., Cheng, C.J., Sun, S.Y.: What influences college students to continue using business simulation games? The Taiwan experience. Computers & Education 53(3), 929–939 (2009)

Tao, Y.-H., Hong, W.J., Yeh, C.R.: Formative evaluation of a monopoly-mechanism online game platform for high usability. In: The 2010 International Conference on e-Commerce, e-Administration, e-Society, e-Education, and e-Technology, Macau, January 25-27 (2010)

Tao, Y.-H., Lee, C.-T.: An executive platform of teaching strategies for peer evaluation. In: ECDL, Shih-Chien University, Taipei, March 17 (2007)

Tsing-Hua Monopoly, Tsing-Hua News (July 8, 2009), `http://www.nthu.edu.tw/allnews/news_content.htm?ID=4698` (accessed February 12, 2012)

Yeh, C.R., Tao, Y.H., Hong, T.P., Lin, W.Y., Chen, P.C., Wu, C.H., Lin, J.: Adapting monopoly as an intelligent learning game for teaching dynamic competitive strategy. In: The 2007 International Joint Conference on e-Commerce, e-Administration, e-Society and e-Education, Hong Kong, August 15-17 (2007)

Serious Gaming: A New Way to Introduce Product Lifecycle Management

Philippe Pernelle[1], Jean-Charles Marty[2], and Thibault Carron[3]

[1] DISP Lab, Université de Lyon
[2] Université de Lyon, CNRS Université Lyon 1, LIRIS, UMR5205, F-69622, France
[3] LIP6, UPMC Paris, F-75005, France

Abstract. Our research work deals with the development of new learning environments, and we are particularly interested in Games-based Learning. We have developed our own learning environment where we apply the metaphor of exploring a virtual 3D world, where each student embarks on a quest in order to collect knowledge related to a learning activity. In this article, we explain how this work has been applied to the industrial domain in order to "conduct the change".

Although the users appreciate this approach, there is an obvious need for information about learners' skills, especially for the teacher. For that purpose, we have equipped the different tools belonging to the learning environment to allow tracing facilities. We can thus update the learners' model when particular events occur, by using both data collected from traces resulting from the (collaborative) learning activity and information collected from the specific business tool integrated in the game.

In this article, we first describe the Game-based learning environment that we have developed. We then focus on how to obtain general information contained in the user profile from basic user's actions (traces). We give details and results on the experiment that we have set up to understand Product Lifecycle Management in such a learning environment. A real experiment has been made at our university in the PLM domain with the help of a company validating the feasibility of the approach.

Keywords: Game Based Learning, Trace observation, Indicators, Process simulation, PLM, collaborative activity.

1 Introduction

Nowadays, compared to traditional teaching methods, Learning Management Systems (LMS) offer functionalities recognized as being valuable from different points of view. For instance, students can learn at their own speed. These environments also allow the teacher to evaluate specific activities in a uniform way. However, although they enable powerful features, some students tend to consider LMS as unexciting (Prensky, 2000).

L. Uden et al. (Eds.): Workshop on LTEC 2012, AISC 173, pp. 89–100.
springerlink.com

Agreeing with Vygotski's school of thought and activity theory (Vygotski 1934), we consider that the social dimension is crucial for the cognitive processes involved in the learning activity. Consequently, the problematic was how to enhance the social dimension in such environments. The emergence of learning games provides a possible answer to this problem and is seen as an evolution of "classical" LMS (Hijon and Carlos, 2006).

As a matter of fact, observing the emergence and success of online multiplayer games with our students, it was decided to experiment our own learning game approach, by developing a new game and using it as a support for some learning sessions. We think the way of acquiring knowledge during a learning session is similar to following an adventure in a Role-Playing Game (RPG). The combination of the two styles is called MMORPG (Massively Multiplayer Online RPG) and offers a good potential for learning (Galarneau and Zibit, 2007; Yu, 2009) reformulated as MMOLE (Massively Multiplayer Online Learning Environment).

We have set a number of experiments that gave interesting results in educational contexts (collaborative aspects, adaptation issues, or immersion topic (Marty and Carron, 2011)). We have recently decided to explore the possibility of using this playful approach to address items useful in the industry. The idea is to use the learning platform to support organizational changes, and more particularly the introduction of Product Lifecycle Management (PLM) systems, in SMEs[1]. (Kadiri et al. 2009) state that the implementation of a PLM system significantly alters the organization of the company, particularly in the context of SMEs. (Individual and collective) resistance to change naturally appears during the start-up of this type of system. We would like to make the user discover by him/herself the gains induced by the change. The learning environment is thus used as a simulation tool, where the user first performs tedious tasks in the old-fashion way (before the introduction of change). S/he is then prepared to receive new ways of working at the end of the session. Furthermore, a user profile can be deduced from the user's possible successes and mistakes (for example: users who haven't chosen the right form to fill in, users who don't enter the correct data, users who haven't applied accurately the business process).

In this article, we first describe the Game-based learning environment that we have developed. We then focus on how to obtain general information contained in the user profile from basic user's actions (traces). We give details and results on the experiment that we have set up to understand Product Lifecycle Management in such a learning environment. Finally we conclude and give future tracks for this research.

2 Learning Adventure Environment

Learning Adventure is a Game-Based Learning Management System representing a 3D environment where the learning session takes place. A particular map (environment with lakes, mountains, hills, and possibly buildings) is dedicated to a specific learning activity, for a specific subject. Each part of the map represents the place where a given (sub) activity can be performed.

[1] Small and Medium Enterprises.

The map topology represents the overall scenario of the learning session, i.e. the sequencing between activities. There are as many regions as actual activities, and the regions are linked together through paths and NPC (Non-Player Character) guards, showing the attainability of an activity from other ones. Similar models that link pedagogical issues with game elements can be found with a more general point of view in (Amory et al., 1999) and more precisely concerning this approach in (Carron, Marty, and Heraud, 2008).

Learning Adventure (L.A.) is based on a role-play approach (Baptista and Vaz de Carvalho, 2008). Players (students or teachers), possibly represented by their own avatars, can move through the environment, performing a sequence of sub activities in order to acquire knowledge. Activities can be carried out in a personal or collaborative way (see (Dillenbourg et al., 1996) for a list of cooperation abilities): one can access knowledge through objects available in the world, via help from the teachers, or through work with other students.

Although such game environments and characteristics are well known from our students the so-called « digital natives », some reminders are always proposed at the beginning of the pedagogical session. This first playable part called the Newbie Park allows us to describe the main functionalities, explain the use of some specific collaborative tools that are present in this game and for a particular learning session. Moreover, as in many collaborative sessions, this part can be seen as a "warm-up activity" in order to get student minds into an adequate "ready to play for learning" state.

The environment is generic in the sense that it is not dedicated to a particular teaching domain. With help from a pedagogical engineer, the teacher adapts the environment before the session by setting pre-requisites between sub activities and by providing different resources (documents, videos, quizzes) linked to the course. Experiments have been set up for learning English as well as Project Management or Object Oriented Concepts in Computer Science. The collaboration takes place in L.A. by constituting some groups of users. The NPCs give objectives to the members of a group and give them access to collaborative tools such as white boards, file boxes or a "collaborative plan elaborator" similar to a structured discussion forum. It is then possible to construct group knowledge with specific tools. Naturally, in order to communicate with other players a chat tool is available.

Learning Adventure thus provides a rich environment to set up diverse learning activities. In industrial contexts, similar difficulties linked to the process change appear. Indeed, some employees need to learn how a new work process can impact the company. We believe that our platform can provide a means for simulation of actions in a company. We have modeled a company building from its plans in order to ensure the players immersion. It is indeed easier for them to find the different places (coffee room, etc…) in the game if they are located at the same places than in the real building. A compass for reaching a specific place is however available in the game (see Fig. 1).

Fig. 1 Learning Adventure screenshots

Playing out a scenario that can actually take place in the company becomes possible. NPCs available in the environment can give instructions (the mission) and hints if needed. The game is configurable and one can easily change the interactions, the layout or the location of the NPCs.

This immersive simulation is helpful for the users to understand why the described process should evolve. It would be even better to know which parts of the process have been difficult to manage for a particular user. The explanation of the introduction of the PLM can then be adapted by focusing on the parts related to the kind of mistakes made by this user. It is our view that the general behavior (and possible mistakes) of a user is part of the user model. This part can be induced from the traces of the activity. In the next chapter, we explain how we can transform this rough information made of basic traces into meaningful indicators.

3 From Traces to Indicators

The tracing activity is an appropriate way to reflect in depth details of the activity and to reveal very accurate hints for the teacher. Unfortunately, traces are very difficult objects to manage and understand. Using traces to understand the global activity raise challenging questions. Is it always possible to collect traces from any activity? How to deal with a huge quantity of information to explore? Are the basic traces at the right level of abstraction to offer a good understanding of the activity?

Due to this complexity, we consider the trace as a research object per se and we agree with the model developed by (Settouti et al., 2009) in the Trace Based Systems (TBS). The idea is to be as generic as possible and to separate the traces (gathered in a data base) from the learning application. In this generic approach, the TBS is in charge of collecting the different traces and of the transformations of traces necessary to interpret the activity.

Collect the traces: The idea is to equip any application (here, the entire Learning Adventure environment) with a tracing possibility. This implies the definition of an API of required basic observations. For instance, in the Learning Adventure environment, actions such as "entering a new zone (workshop)" or "the teacher is helping someone" may be traced and thus collected by specific elementary probes.

Basically, in our environment, we defined 17 elementary probes that can be flagged at any moment by any client of our application. Some examples of probes are given in table 1, each probe containing parameters and having a particular aim in order to fulfill a specific category of awareness, which is not often addressed (see (Gutwin and Greenberg, 2002) for awareness definition).

Another item to be considered is the way of tracing an external component that must be integrated in the pedagogical platform. It is the case in our application, where we have integrated business tools interacting with the Learning Adventure Environment. These tools are also equipped with specific probes. For example, the actions done through the interface are collected in the game. We are thus able to know whether a wrong manipulation occurred in the business tool, if a user provided a positive or a negative comment, if someone asked for explanation, etc.

Transform the traces: The transformations of traces are mandatory to understand the activity of the learning session. To address the problems of the quantity of traces and incorrect level of granularity, we have considered two kinds of transformations: the filtering transformations and the abstraction transformations.

The filtering transformation is quite simple. It acts as a request in a database where the only traces that will be considered are linked with a particular view of the activity (e.g. communication traces or traces concerning User X).

For a teacher, the expectations concerning the perception through the system are somewhat difficult to express. The level of what needs to be perceived may vary, as it is also the case in traditional teaching: a teacher may want to observe basic facts (e.g. who starts a new activity) or more abstract facts (e.g. who regularly cooperates before answering a quiz properly). The API provides the users with elementary probes. They are thus useful for observing basic facts. However, they may not be helpful enough when the level of abstraction needed is higher.

Table 1 Some Awareness Probes (among the 17 ones)

Probe Name	Parameters	Awareness category
WorkshopArriving	<UserName>, <WorkshopName>	Group-structural awareness
WorkshopLeaving	<UserName>, <WorkshopName>	Group-structural awareness
ChatContentListening	<UserName>, <ChannelName>, <Sent-Message>	Social awareness
GroupSplitting	<GroupName>, <GroupSplitter>, <User-Name1>, <UserName2>, ...	Group-structural awareness
StudentConnecting	<UserName>	Informal awareness
TeacherHelpCalling	<UserName>, <TeacherName>	Social awareness
...		

For instance, being aware only of a student consulting a help file can be not very meaningful. But if the same observation occurs just after s/he has given a wrong answer and then followed that with a success in the same activity, the teacher may be reassured as to the usefulness of the help file related to the activity.

The combination of these three indicators (simple probes) allows to create a complex probe and thus to provide a higher-level explanation about the on-going activity.

As stated previously, 17 basic indicators (simple probes) have been extracted from your learning environments and 3 operators have been proposed: AND, OR, THEN to combine one probe with another. All these indicators may be combined with each other to define new complex probes and their representation with an administration tool. The new indicator is available in the educational platform, with the same properties as the basic probes. In particular, the new ones can be reused to create a more complex probe or indicator. Thanks to this mechanism, the set of indicators naturally increases with the help of the users, guided by the needs of observation in the platform.

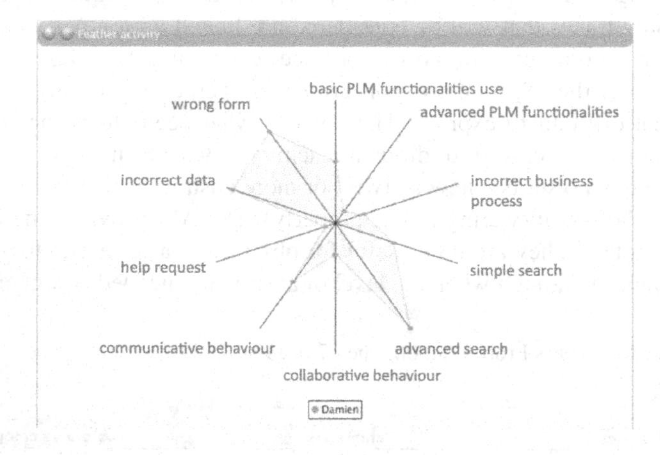

Fig. 2 Individual view of user's actions in the learning environment

This way of transforming the traces allows the teacher to obtain more abstract indicators, close to the application understanding, and easy to display for further analysis. Figures 2 and 3 are examples of results of such transformations. Figure 2 is dedicated to a classification of a single user's actions while figure 3 refers to a collaborative activity, presenting an indicator of collaboration (Gendron, Carron, and Marty, 2008).We now come back to our example in industrial domain and explain how we apply this work to the problem of the Product Lifecycle Management (PLM) in the Learning Adventure Environment. This work has been carried out in a project called PEGASE.

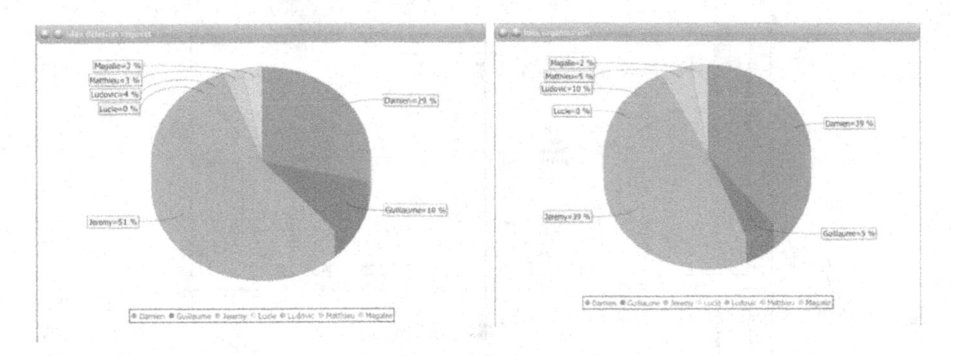

Fig. 3 Collective view of users' participation to a collaborative action

4 Scenarisation and Application to Product Lifecycle Management

PLM systems affect the business practices of the organization. Changing the organization creates difficulties in working, regardless of any technological problems. According to (Kadiri et al., 2009), implementation of PLM systems implies difficulties or rejection 45 times out of 100. The causes of these difficulties are diverse (system failures, poor ergonomics, etc.) but often result from employees' reluctance to use the system. Finally, implementation of a PLM system significantly alters the organization of the company, particularly in the context of SMEs. Resistance to change (individual and collective) appears naturally during the start-up of this kind of systems. Starting from this statement, a challenging issue consists in delivering rich content to companies with an attractive environment. For that purpose, the gamification process, i.e. the transformation of common business processes into one or several game scenarios, is our first objective.

In order to make people sensitive to PLM, we have chosen a simple industrial process (purchase order) described in Figure 4. Without a PLM system, this process is achieved through traditional activities where the risk of error, as well as the tedious tasks involved, should be considered.

A simulation of this process is offered in L.A. without any PLM system. Figure 5 describes some examples of actions (discussion with colleagues and watching a training presentation, collecting documents in order to complete and visualize the tasks in the process, visualization of the order form).

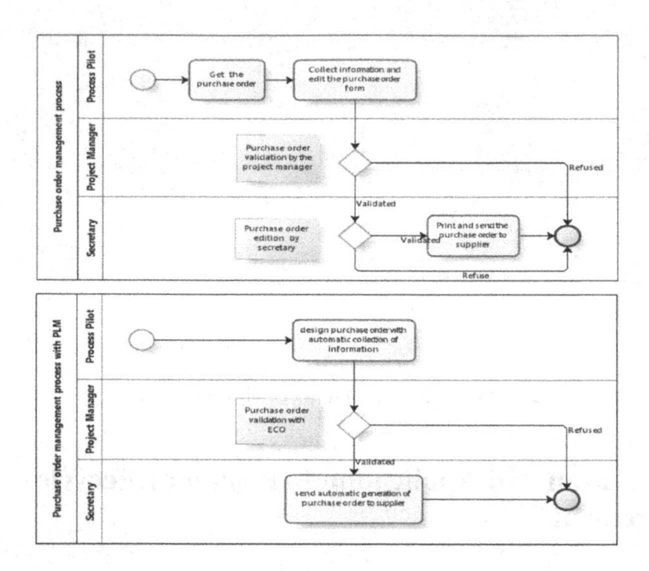

Fig. 4 Purchase order management process with and without PLM

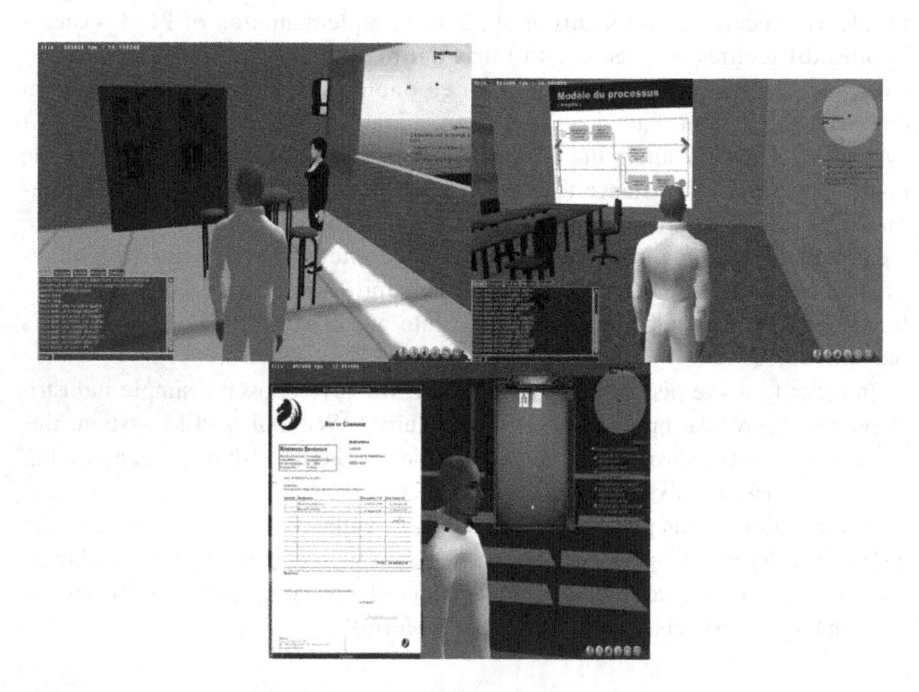

Fig. 5 Screenshots of different steps of the scenario

Once the process has been carried out without a PLM system, mistakes made by the user are displayed. A short training session on possible solutions to solve their mistakes with the Audros PLM system is then proposed. The process is therefore performed in the game with a connection to the Audros PLM system.

The aim of the modeled scenarios is to understand the value of PLM but also to evaluate specifically some skills concerning the user in this domain. The game is thus enriched with indicators allowing updating the user profiles.

As explained in section 3, the environment is able to collect hints on activities achieved by the user and to present them to the supervisor as graphical indicators (Gendron et al., 2011).

5 Experiment Description and Results

This experiment was carried out in 2011 in the University of Lyon with co-located settings (see Fig. 6). During the experiment, one group of twelve students with their teacher was present in the classroom equipped with 12 computers. Concerning the social presence perception (de Kort and Ijsselsteijn, 2008), players were oriented away from each other, limiting mutual eye contact, natural reciprocation of approach or avoidance cues and mirroring. The students were between 24 and 50 years old (PLM Master) and for most of all familiar with computer use. Each student accessed the virtual environment through his/her workstation, and had a personal (adapted) view on the world. These students used the environment for approximately one hour and a half. They were explicitly allowed to communicate through the chat tool provided with the system and were warned that they would be observed regarding the use of the system. Two observers were present in the classroom.

Fig. 6 Conditions of the experiment

As explained before, the course was dedicated to the understanding of the PLM problematic. The learning content dealt with "PLM elementary concepts": the session was split into two parts: experimenting without (long and tedious) and with a PLM system. The solution (final document) had to be approved by a PLM system expert who was online via the PLM business tool and was receiving each document produced. The aim of the session was to assess the knowledge and know-how of the students about the objectives of a PLM system. A story guided the knowledge quest thanks to metaphors. Indeed, the challenge is encouraged through NPC who propose a coherent contest. Immersion is reinforced when the users' actions have a direct impact on the objects of the world. Finally, the teacher was also present in the game via an avatar: it was possible to chat with him, to ask for help for example.

Thanks to traces and indicators, correct actions are automatically rewarded with the relevant skill level up. The learner's user model is consequently updated in the game and the teacher or the PLM expert is able to take the right decisions in order to improve the learning aspects.

Concerning this experiment, two ways of evaluation were chosen:

- Quantitative, thanks to collaborative indicators elaborated with traces left by the users when collaborating,
- Qualitative, with live questions at the end of the session and explicit feedback from the teacher/PLM expert.

At the end of the experiment, the students were asked to answer questions in order to give feedback about their feelings concerning their work session. The questions (ranking and open-ended questions) evaluated aspects relating to several parts of the learning game (pedagogical content, business concepts, affordance of the objects or tools, scenario-story, immersion, collaborative activities and specifically collaborative tools, and user model evolution). The final question let the students propose improvements concerning pros and cons of the game. For example, they found that the chat must be improved (not receiving all messages but contextualized "near-me" messages).

The initial objective concerning PLM problematic perception was reached. All students would rather be working with this environment than in a more conventional and practical way, and more generally were very enthusiastic about this kind of experiment.

From the teacher's point of view, in the light of previous experiments (Bisognin et al., 2010), it was mandatory to have tools supporting him in the monitoring task with the help of an updated user model for each student. As explained before, several indicators or observation widgets (see fig. 2 and fig. 3) were therefore set up in our environments in order to meet this requirement. The results were satisfying but the set-up of such tools and experiments is extremely time-consuming for the teacher.

6 Conclusion

In this article, we have illustrated a way of making the introduction of new work processes in companies more attractive. A game environment is the main support tool proposed to motivate people to discover the assets of a new process. Immersion is an essential characteristic in the simulation environment that offers tracing facilities. This feature allows the teacher or the PLM expert to be aware of the main difficulties encountered by the users while doing their simulation sessions. An experiment to conduct the change in industry has been set up with our students in the Learning Adventure environment.

Currently, users are all present in the same environment and may communicate via chat. Scenarios focusing on collaborative aspects are currently implemented regarding product non-conformity detection in PLM. In fact, the entire infrastructure is ready to host more elaborated scenarios where a group of players cooperate in a company scenario (with different identified roles). As stated previously, this environment is still being developed and we thus aim at proposing both specific collaborative tools and facilities to help the teacher regulate a learning session. Moreover, in perspective, we would like to observe how teamwork could be self-regulated thanks to specific functionalities of collaborative tools. We have shown an example based on several players interacting and communicating in the same environment but with the use of real teams and roles, we may also imagine self-regulation, team-regulation or auto regulation by the use of specific rules as in Artificial Intelligence. Self-regulation, co-regulation and socially-shared regulation are precisely described in (Hadwin et al., 2010).

Naturally, some drawbacks remain: we must admit that it is very difficult for the teacher to be present in the game, help the students and regulate the session even with these specific tools. At the moment, indicators are not very well integrated because they are very time-consuming to develop. We believe that we can develop some specific generic indicators dedicated to classical fields of a domain. An interesting perspective would be to develop and propose basic regulation actions (such as "play specific PLM video", "propose new adapted content or new adapted scenario" or "enable/disable such facility/ies for this student") directly accessible through the indicators.

Finally, in another project called SLI (Serious Lab for Innovation), also supported by the French Ministry for the Economy, Industry and Employment, we also applied this work to another domain: Innovation in industry. Naturally, other scenarios were designed, developed and proposed for that purpose. Some qualitative results concerning concepts and methodology about innovation are still under processing.

From a more general point of view, "conduct the change" is a key point both for industrials and students that will experiment such problems later. The instructions for the introduction of the change can thus be customized according to this profile (Brusilowski, 2001). Generic serious games approach will help us to bridge the gap between education courses (theoretical approach thanks to pedagogical resources left in the game) and concrete experiment in professional context thanks to the integration of real business tool (access to interfaces via web services).

Acknowledgments. We would like to thank the French Ministry for the Economy, Industry and Employment (DGCIS) for the support in the PEGASE project. We would like also to thank G. Dalla Costa, J. Depoil, L. Kepka, and L. Michea for their great help in developing the Learning Adventure Platform.

References

Amory, A., Naicker, K., Vincent, J., Adams, C.: The use of computer games as an educational tool: Identification of Appropriate Game Types and Game Elements. British Journal of Educational Technology 30(4), 311–321 (1999)

Baptista, R., Vaz de Carvalho, C.: Funchal 500 years: Learning Through Role Play Games. In: Proceedings of ECGBL 2008, Barcelona, Spain (2008)
Bisognin, L., Carron, T., Marty, J.-C.: Learning games factory: Construction of learning games using a component-based approach. In: Proceedings of European Conference on Games Based Learning (ECGBL), Copenhague, Danemark (2010)
Brusilovsky, P.: Adaptive hypermedia. User Modeling and User Adapted Interaction, Ten Year Anniversary Issue (Kobsa, A. ed.) 11(1/2), 87–110 (2001)
Carron, T., Marty, J.-C., Heraud, J.-M.: Teaching with Game Based Learning Management Systems: Exploring and observing a pedagogical dungeon. Simulation & Gaming Special issue on eGames and Adaptive eLearning (2008)
De Kort, Y.A.W., Ijsselsteijn, W.A.: People, Places, and Play: Player Experience in a Socio-spatial Context. Computers in Entertainment 6(2) (2008)
Dillenbourg, P., Baker, M., Blaye, A., O'Malley, C.: The evolution of research on collaborative learning. Learning in Humans and Machine: Towards an Interdisciplinary Learning Science, 189–211 (1996)
Galarneau, L., Zibit, M.: Online Game for 21st Century Skills. In: Gibson, D., Aldrich, C., Prensky, M. (eds.) Games and Simulations in Online Learning: Research and Development Frameworks, pp. 59–88. Information Science Publishing, Hersey (2007)
Gendron, E., Carron, T., Marty, J.-C.: Collaborative indicators in Learning Games: an immersive factor. In: 2nd European Conference on Games Based Learning, Barcelona, Spain (October 2008)
Gendron, E., Pourroy, F., Carron, T., Marty, J.C.: Towards a structured approach to the definition of indicators for collaborative activities in engineering design. Journal of Engineering Design, 1–31 (May 2011)
Gutwin, C., Greenberg, S.: A descriptive Framwework of workspace awareness for real-time groupware. In: Computer Supported Cooperative Work, vol. (11), pp. 411–446. Kluwer Academic Publishers (2002)
Hadwin, A.F., Oshige, M., Gress, C., Winne, P.H.: Innovative ways for using gStudy ot orchestrate and research social aspects of self-regulated learning. Computers in Human Behavior, Advancing Educational Research on Computer-Supported Collaborative Learning (CSCL) through the use of gStudy CSCL Tools 26(5), 794–805 (2010)
Hijon, R., Carlos, R.: E-learning platforms analysis and development of students tracking functionality. In: Proceedings of the 18th World Conference on Educational Multimedia, Hypermedia & Telecomunications, pp. 2823–2828 (2006)
Kadiri, S., Pernelle, P., Delattre, M., Bouras, A.: Current situation of PLM systems in SME/SMI: Survey's results and analysis. In: PLM 2009, Bath, England (2009)
Marty, J.-C., Carron, T.: Observation of collaborative activities in a game-based learning platform. Transactions on Learning Technologies (TLT) 4(1), 98–110 (2011)
Prensky, M.: Digital Game-Based Learning. MacGraw Hill, New York (2000)
Settouti, L.S., Prié, Y., Marty, J.C., Mille, A.: A Trace-Based System for Technology-Enhanced Learning Systems Personalisation. In: Proc. of the 9th IEEE Int. Conference on Advanced Learning Technologies (ICALT), Riga, Leetonia, pp. 93–97 (2009)
Vygotski, L.S.: Language and Thought, Gosizdat, Moscow (1934)
Yu, T.W.: Learning in the Virtual World: the Pedagogical Potentials of Massively Multiplayer Online Role Playing Games. International Education Studies 2(1) (2009)

A Study on Factors Influencing Students' Participation in Skill Certification Test

Chin-Wen Liao, Chen-Jung Lai, Fang-Pin Lai, and Li-Chu Tien

Department of Industrial Education and Technology,
National Chunghua University of Education. No.2, Shi-da road,
Chunghua City, 500, Taiwan, R.O.C.
tcwliao@cc.ncue.edu.tw, {tcivs74,m9316621}@yahoo.com.tw,
lion0829@gmail.com

Abstract. The objective of this study was to investigate factors influencing students' participation in Level II Skill Certification. This research was carried out using a survey questionnaire approach, in which a questionnaire *Factors Influencing Students' Participation in Level II Skill Certification* was developed based on literature review results. A total of 1,010 copies of questionnaire were distributed among graduating students picked by the stratified random sampling method. A total of 848 copies of questionnaire were returned and valid. The collected data were statistically analyzed by the independent samples *t*-test, one-way ANOVA, and Scheffe's test for multiple comparisons (post-hoc). The results are:

1. The factor *"Parents' Expectations"* was the most influential factor for students participating in the Level II Technician Skill Certification.
2. Private school students were more affected than public school students were.
3. The factor *"Achievement Motivation"* was more influential on students planning for continuing their education than on students planning for going into the job market.

Keywords: Electronics and electrical engineering, Level II technicians, Influential factors.

1 Introduction

Over the last ten years, the fast political, economical, social and educational advancement has pushed the industries toward high technology, informatization, and automation.

Technological and occupational education and skill certification are closely linked. Professional skill certificates are beneficial for students wishing either to continue their education or to go into the job market (Cheng, 2009). Because of the trend of applying occupational certification system to all industries in advanced countries, medium-level vocational schools have started to put more and more emphasis on occupational certification training. As a result, promoting technician skill certification through technological education is an important education policy at this moment (Chang & Jao, 2010).

L. Uden et al. (Eds.): Workshop on LTEC 2012, AISC 173, pp. 101–109.
springerlink.com

For implementing the professional certification system, the Ministry of Education (2009) has pointed out that without affecting the regular course work, vocational high school students are encouraged to participate in *Level I Technician Skill Certification Test* in the school to elevate their skill proficiency and get ready for the job market. Once the students have obtained the *Level I Certification*, they can also pursue *Level II Technician Skill Certification* to advance their professional skills and acquire more vocational advantage.

Having taught electronics and electric engineering related professional courses and internship at vocational high schools for many years, the researcher is very familiar with the current education environment and situation. His objective for this study was to investigate factors affecting students' participation in *Level II Technician Skill Certification Test*. The research finding can be used for policy-making by government agencies in the future.

2 Problem Description and Background

2.1 The Meaning and Classification of Technician Skill Certification Test

Certification and license are not equivalent terms. Certification is usually issued by a host agency. It certifies an individual's capability but holds no legal effect. License, on the other hand, is issued by the authority of a specific profession and holds specific legal effects (Tan, 2007).

The Technician Skill Certification System currently implemented in Taiwan has three levels: Level I, II and III. In this case, the higher the level suggests a greater aptitude in terms of professional knowledge and proficiency. For technicians at Level II, they should possess the ability to apply and analyze knowledge relevant to their profession and be familiar with the operation, technical design, planning and instruction at a journeyman level.

2.2 Factors Affecting Students' Participation in the Technician Skill Certification

Students' willingness to participate in the Technician Skill Certification and factors affecting their participation are associated with various social science theories. The three most important factors are personal achievement motivation, expectations from parents, and expectations from teachers (Hsiao, 1994).

Students' behavior, attitude and sense of worth are profoundly influenced by their peers (Lin, 1980). From a cultural perspective, school administrative support, a type of secondary culture, definitely has a certain degree of influences on students' learning and performance at school. Taken together, this study therefore divided factors affecting students' participation in *Level II Technician Skill Certification* into six aspects.

(1) Achievement motivation

Achievement motivation is an internal force prompting individuals to pursue self-achievement (Chang, 1999)

(2) Parents' expectations

Expectation is about a conscious or subconscious value created by individuals on themselves or by others (Finn, 1972).

(3) Teachers' expectations

It is about the value and expectations from teachers on each student based on the student's personality, attitude and sense of worth (Cheng, 2000).

(4) School administrative support

School administration is processes carried out by school agencies for handling assorted school-related matters, including personnel, finance, schedules and business, appropriately and effectively.

(5) Peer influences

Davis (1989) considered that there is an interactive peer influence among adolescents. Huang (1990) found that during the leisure time of adolescents who are enrolled students, their partners in leisure time activities are more likely to be their peers rather than family members or others.

(6) Effectiveness of the Technician Certification

According to Selection and Recommendation Admission Regulations Regarding Technically and Artistically Talented Students at Middle/ Senior High Schools & Institutions of High Learning, individuals with a Level II Technician Skill Certificate are given an additional 15% of the final admission score.

According to *Appraisal of Self-Directed Academic Achievement* (Ministry of Education, 2010), individuals with a *Level II Technician Skill Certificate* can apply for the appraisal for junior college level of academic achievement.

According to Article 3 of Regulations Regarding the Selection and Appointment of Specialized Technical Instructors at Junior Colleges (Ministry of Education, 2010), individuals graduated from junior colleges or above, with a Level II Technician Skill Certificate, and have at least four years of work experience in a field related to the subject of the appointment are eligible to be appointed as lecturer-level specialized technical instructors.

According to *Regulations Regarding the Review and Evaluation for Specialized Technical Instructors at Vocational Schools* (Ministry of Education, 2010), individuals registering for this review and evaluation process have to be junior college graduates or above, hold a *Level II Technician Skill Certification* for the interested subject, and have at least two years of work experience where relevant professional skills have been applied.

Since 2006, the Ministry of National Defense has offered military officers and solders with a *Technician Skill Certificate* from R.O.C. the priority for placement based on their specialty. The Council of Indigenous People of the Executive Yuan (2010) offers aboriginal people with a *Level II Technician Skill Certificate* a reward of 5.000 NTD.

According to the above regulations and laws, individuals with *Technician Skill Certificates* have big and tangible advantages, regardless of whether they want to advance their education or to go into the job market. These advantages should be influential enough on affecting vocational high school students' participation in *Level II Technician Skill Certification.*

3 Research Design and Implementation

3.1 Research Framework

The research framework, constructed based on the research objectives and results from the literature review, is presented in Figure 1.

3.2 Variance Test

3.2.1 Research Subjects

The study subjects were Grade 12 graduating students (N = 4,757). Using the stratified random sampling approach, a total of 1,010 copies of questionnaire were mailed to students randomly sampled from the pool. After eliminating eighty-five invalid copies of questionnaire, a total of 848 copies of questionnaire were valid (83.96%). See Table 1.

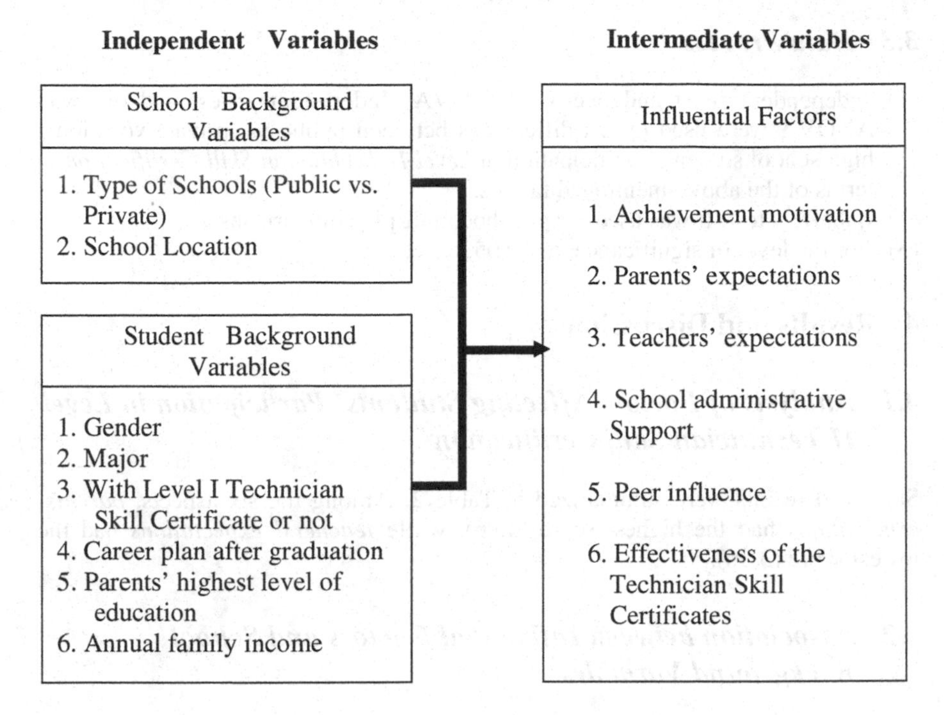

Fig. 1 Research framework

Note:
This sign ⟶ denotes influences from background variables on students' participation in *Level II Technician Skill Certification.*

Table 1 Data distribution of the study sample

School Background Variables	Factors	No. of People	%
Private vs. Public School	Public Schools	339	39.98
	Private Schools	509	60.02

Note : N = 848

3.2.2 Research Tools

The questionnaire used in this study, Questionnaire for Students' Level II Technician Skill Certification Participation, was developed by the researcher.

3.3 Data Process

(1) Independent *t*-test and one-way ANOVA: Independent *t*-test and one-way ANOVA were used to test differences between public vs. private vocational high school students' participation in *Level II Technician Skill Certification* in terms of the above-mentioned factors.
(2) Scheffe's test: It was used for post-hoc multiple comparisons
(3) For the level of significance, $\alpha = 0.05$.

4 Results and Discussion

4.1 Analysis of Factors Affecting Students' Participation in Level II Technician Skill Certification

Statistical results were summarized in Table 2. Among the six aspects, *parents' expectations* had the highest score (3.96), while *teachers' expectations* had the lowest score (3.56).

4.2 Association between Influential Factors and School Background Variables

Statistical results concerning school background variables and influential factors were summarized in Table 3:

5 Conclusion and Suggestions

5.1 Conclusion

(1) The factor "Parents' Expectations" was the most influential factor affecting students' participation in Level II Technician Skill Certification.
(2) *For* Level II Technician Skill Certification *participation, private school students were more influenced by* parents' expectations, teachers' expectations, peer influences *and* school administrative support *than public school students were.*
(3) *For* Level II Technician Skill Certification *participation, students planning to advance their education after graduation were more influenced by the* achievement motivation *than those planning to join the job market were*

5.2 Suggestions

(1) Schools should pay attention to the followings: ①Actively promote information related to *Level II Technician Skill Certification.*② Provide good hardware facilities. ③Open *Level II Technician Skill Certification* related training courses.

(2) Teachers should pay attention to the followings: ① They are encouraged to obtain the *Level II Technician Skill Certificate* for the course they teach. .② They can share *Level II Technician Skill Certification* related information in the class with their students.
(3) For the future research, researchers can interview randomly selected students or carry out experimental research to further investigate those influential factors as well as to elevate the reliability.

Table 2 Combined analysis of factors affecting students' participation in *Level II Technician Skill Certification*

Aspects	Mean Score	SD	Rank	Level of Influence	Semi-Structure Cognitive Percentage
Achievement motivation	3.89	4.91	2	Highly influential	75.12
Parents' expectations	3.96	3.20	1	Highly influential	76.65
Teachers' expectations	3.56	4.41	6	Highly influential	68.86
Peer influence	3.65	3.93	5	Highly influential	70.15
School administrative support	3.70	4.67	4	Highly influential	72.56
Technician certification effectiveness	3.86	3.68	3	Highly influential	74.33
Overall	3.77	19.63		Highly influential	

Note : N=848

Table 3 Combined analysis of school background variables and factors affecting students' participation in *Level II Technician Skill Certification*

Aspect	Public School vs. Private School
Achievement motivation	
Parents' expectations	(2)>(1)
Teachers' expectations	(2)>(1)
Peer influence	(2)>(1)
School administrative support	(2)>(1)
Technician certification effectiveness	
Overall	(2)>(1)

References

Chang, C.C., Jao, T.C.: Consideration of Development of the Certification System for Technical and Vocational Education. Journal of Educational Resources and Research 93, 15–30 (2010)

Cheng, L.C.: Problems related to vocational high school students' Level 1 Skill Certification Test. Taipei County Education 66, 78–80 (2009)

Davis, F.D.: Perceived usefulness, perceived ease of use, and user acceptance of information technology. MIS Quarterly 13(3), 319–342 (1989)

Dyson, L.: Training and vocational education programs are revised. Rural Conditions and Trends 10, 36–39 (1999)

Executive Yuan, Incentives for Aboriginal People Obtaining Technician Skill Certification. From the website of the Council of Indigenous People, Executive Yuan (October 2, 2010), `http://law.apc.gov.tw/LawContent.aspx?id=FL039983&KeyWord=%e4%b9%99%e7%b4%9a`

Executive Yuan, Occupational Training Law. From the website of the Council of Labor Affairs of the Executive Yuan (September 12, 2010), `http://laws.cla.gov.tw/Chi/FLAW/FLAWDAT0201.asp`

Executive Yuan, Regulations Regarding the Registration for the Appointment of Specialized Technical Instructors at Vocational Schools. From the website of the Department of Personnel, Ministry of Education (October 2, 2010), `http://www.edu.tw/human-affair/content.aspx?site_content_sn=8753`

Finn, J.D.: Expectations and the educational environment. Review of Educational Research 42(3), 387–410 (1972)

Greene, K.V., Kang, B.G.: The effect of public and private competition on high school outputs in New York State. Economics of Education Review 23, 491–506 (2007)

Hsiao, Y.D.: Central Regional High School graduates of post industrial Branch to participate in the school skills testing willings and its related factors. Master's thesis of Industrial Education, National Changhua University, Changhua County (1994)

Hsu, H.G., Rnu, D.C.: A study of the strategies for promoting vocational senior high school into community college. Paper Presented at the International Vocational Education and Training Association Annual Conference, Sydney, Australia, August 11-13 (1999)

Kagaari, J.R.K.: Evaluation of the effects of vocational choice and practical training on students' employability. Journal of European Industrial Training 31(6), 449–471 (2007)

Lin, C.C.: Graduates entering employment will affect factors. Taiwan Normal University Institute of Educational Research Bulletin 22, 129–188 (1980)

Ministry of Education, Appraisal of Self-Directed Academic Achievement. From the website of Laws and Regulation Database (October 2, 2010), `http://law.moj.gov.tw/LawClass/LawContent.aspx?PCODE=H0080010`

Ministry of Education, Selection and Recommendation Admission Regulations Regarding Technically and Artistically Talented Students at Middle/ Senior High Schools & Institutions of High Learning. From the website of the Executive Yuan Gazette Online (October 2, 2010), `http://gazette.nat.gov.tw/EG_FileManager/eguploadpub/eg012222/ch05/type2/gov40/num11/Eg.htm`

Ministry of Education, From the website of Laws and Regulation Database (October 2, 2010),
http://law.moj.gov.tw/LawClass/LawAll.aspx?PCode=H0040004

Ministry of National Defense, Act of Assignment for Officers and Noncommissioned Officers of the Armed Forces. From the website of the Department of Defense National Defense Laws and Regulations Database (October 2, 2010),
http://law.mnd.gov.tw/scp/Query1A.asp?no=1A003000027&K1=technician

Rau, D.C.: Study of constructing instructional indicators for the electric and electronic professional subjects teacher in vocational senior high school. Journal of Technological and Vocational Education 2, 99–108 (1999)

Youshihisa, K.: Person, symbol, and sociality: Towards a social psychology of cultural dynamics. Journal of Research in Personality 38, 38–52 (2007)

A Study of the Effectiveness of Using Blog as a Cooperative Learning Platform for Accounting Skill Certification Test

Chin-Wen Liao, Li-Chu Tien, Sho-Yen Lin, and Hsuan-Lien Chen

Department of Industrial Education and Technology,
National Chunghua University of Education. No.2, Shi-da Road, Changhua City 500, Taiwan, R.O.C.
tcwliao@cc.ncue.edu.tw, m9316621@yahoo.com.tw,
linshoyen@gmail.com, hsuan.lien@msa.hinet.net

Abstract. This study aims to investigate the effectiveness of cooperative learning for accounting skill certification test by using blog as a platform. The investigation applied quasi-experimental approach. The statistical sampling was from the second year students in the information operation department of a local private vocational academy. Students were S-type grouped into two groups, namely, experimental group and controlled group based on their academic performances. The former used blog as a platform of both teaching and cooperative web learning; the later only used blog as a teaching platform. The learning subject that is the accounting skill certification test related material is identical for both groups. After 18 weeks of application, the results indicated that the students in experimental group had better learning effectiveness over those in the controlled group with higher degree of satisfaction.

Keywords: cooperative e-learning, teaching blog, learning achievement.

1 Introduction

In this era of information economy, new information technologies continue to emerge, and every country is striving toward the goals of technological innovation and globalization (Cheng and Chen,2008). In this age of information explosion, students' demands on information grow rapidly; therefore, the linear and verbal teaching process is changed into applying information to teaching strategies. A teaching blog provides curriculum related information and teaching material downloads to students and information sharing among students and other instructors (Black, 2007). Students can also share information with instructors and classmates after class through the blog, which motivates students to actively participate in the discussion (Black, 2007). Therefore, instructors can use the online teaching blog as a platform for student's cooperative e-learning, which enables students' self-learning through the web-based teaching platform at any time. Because the learning strategies and methods using a teaching blog can be

L. Uden et al. (Eds.): Workshop on LTEC 2012, AISC 173, pp. 111–120.
springerlink.com

used for structural, unstructured, or mixed collective learning, without the limitation of class duration and location and space of a classroom, students can learn in any place (Sun &Lin, 2007).

Although a teaching blog platform is an online personal blog, it allows students to participate in the learning process, inspires students' creativity and problem solving capability, and encourages valuable information accumulation and sharing. Using the teaching platform, instructors can have more interactions and discussions with students, understand students' questions and learning difficulties, and provide after school education, which will improve students' learning achievement. Gallagher, Jamey (2010) stated that a blog is a new communication environment to enhance students' learning motivation and learning achievement. In the recent years, national and international studies have proven that a teaching blog platform has positive effects on student satisfaction and learning achievement (Arslan, Receps & Sahin-Kizil, Aysel, Gallagher, Jamey, 2010; Oikonomidoy, Eleni ,2009; Chuang & Shen, 2008; Jheng & Lai, 2007). Philip, Robyn & Nicholls, Jennifer (2009) pointed out that blogs (online diary) are increasingly being used by higher education to evaluate students' cooperative learning and feedbacks. Hou, Chang & Sung (2009) also suggested that a teaching blog, as an interactive media over internet, is the best online information-building platform for interactions among students or between students and instructors.

Accounting involves many difficult concepts and terminologies, such as the definitions of retail business and service industry and the difference of their operation ideals and models. In order to comprehend the abstract, complicated, and difficult accounting concepts and terminologies, the functions of "group discussion," "feedback," and "questioning and experience sharing" on the teaching blog can be used for students' cooperative learning and extensive discussion in groups and for instructors' explanation. Instructors then post the homework on the teaching blog based on the exam type. Students can complete and send back their homework through "feedback" function. By this way, the abstract and incomprehensible concepts can be materialized and simplified to motivate students to study and achieve their learning goals. Therefore, teaching blog platforms are very useful for students' cooperative e-learning over. Many prior studies had proven that cooperative e-learning improves students' learning performance (Chiu, 2004; Hsu, 2003). The national and international literatures also pointed out that the application of teaching instruments on blog platforms for science educations, such as mathematics, physics, biology, and medicine, yields good results (Wang & Wang, 2004; Mei-Chung Lin, 2009; Meltem Huri Baturay; Bay & Omer Faruk, 2010; Wang & Ma,2008). However, no study has been conducted to investigate the impact of cooperative e-learning using teaching blog platforms on students' accounting skill test result..

Based on the above arguments, the objectives of this study are as follows:

(1)Verify whether the application of teaching blog platforms to students' web-based cooperative learning is helpful for achieving high scores in the accounting technician examination.

(2) Examine students' perceptions and attitude toward accounting cooperative e-learning using teaching blog platforms.

2 Literatures

2.1 Teaching Blogs

The word blog or "web blog" was invented by Jorn Barge in 1997. A blog can be used to write a personal diary, express individual idea, or writing memos by group members. The blog software provides the functions of user defined organizing settings and automatically saving the files, which are excellent for memorandum and data management and storage. Harder and Reichardt (2003) pointed out that blogs include latest news, interesting website links, comments, information sharing, etc.. Conhaim (2002) thought that a blog, which facilitates individuals' knowledge assimilation, self-expression, and self-creation of a trend, is a brand new communication channel and a broadcasting model. The advantages of powerful, quick news broadcasting and self-learning features enable the rapid growth of blogs. A teaching blog constructed by an instructor using a blog platform allows students to study actively, interact with others, and learn from others' experiences (Liu, 2009). This study used the blog registered on PIXNET as a platform of cooperative e-learning to provide students a cooperative e-learning tool for after school studying.

2.2 Theories of Cooperative E-Learning

The theories of cooperative e-learning are mainly divided into "cognitive theory" and "affective theory" (Sun & Lin, 2007):

1. Cognitive theory
(1). Cognitive elaboration theory
The cognitive elaboration theory suggests that in order to achieve optimal learning results, students must create associations between new information and existing information in the long-term memory, so that the new information can be stored in the long-term memory effectively.
(2). Situated cognition
This theory claims that information is distributed all over the world, possessed by different people, located in the tools or books people use, and stored in every professional community of every sector. Therefore, learning occurs in any situation.
(3). Social cognitive theory
This theory focuses on "social environment" and "inner cognitive activities". It is believed that a learning process involves not only a conceptual change or learning new skills and strategies for individual student, but also the interactions among peers in social situations, such as inquiry and imitation from each other, which is required to motivate the learning and strengthen the learning result.

2. Affective theory:
(1) Motivation:
There are three motivation components: value component, expectancy component, and affective component. All of these are effective predicators of learning behaviors.
(2). Self regulated learning
For the subjects in which students are interested, students will search for learning materials, compile and analyze the questions, discuss the questions with others, ask for instructors' help, or change the problem solving approach, without instructors pushing them or the pressure of exams.

3.Examination of accounting affair skills
The examination of accounting technician has two levels, level B and level C, with the test subjects of general subjects and special subjects. Those who score sixty or above on both subjects can obtain the qualification of accounting technicians. The general test subjects include basic accounting knowledge, and special test subjects include the practice of creating accounting vouchers, bookkeeping, and preparing accounting statements. The exam for level C tests the basic skills of the six major accounting procedures and the related knowledge. The exam for level B tests the knowledge about accounting documents, account structure, partnership and corporate accounting, and evaluation of assets and debts. The pre-test questions were selected in this study from the question database built by the Bureau of Employment and Vocational Training, Council of Labor Affairs, Executive Yuan for regulating the examination of level C accounting technician. The scores of this examination are used to evaluate students' study performance.

3 Study Design

The unequal quasi experimental method was adopted in this study. The s-shape grouping strategy was used to group students according to their accounting scores of previous semester. An experiment for validating this study is arranged in two groups: an experimental group of twenty-five people and a controlling group of 24 people from the data processing classes' students of 11th grade in a private vocational high school. Both groups took four experimental classes every week for eighteen weeks. There are two independent variables: one is instructed by the cooperative e-learning method using the accounting teaching blog platform; the other is using the traditional learning method through the accounting teaching blog platform. The learning satisfaction and learning achievement are evaluated for each variable. Before the experiment, both groups have learned accounting and have the same accounting knowledge. In the study design, both groups should receive a pre-test before taking the class, using the questions from the database of "examination of level C accounting technician." The result of pre-test was collected (O1) . The experimental group is instructed by the cooperative e-learning method using the accounting teaching blog platform (X1). The controlling group is using the traditional learning method through the accounting teaching blog platform (X2). After the experiment, both groups took the examination of level C accounting technician provided by the Council of Labor Affairs. The test result was then collected(O_2). The Kuder-Richardson reliability verification was performed for the questions in the pre-test and the post-test. The design model of unequal controlling group is as follows:

$$\begin{array}{ccc} O_1 & X_1 & O_2 \\ O_1 & X_2 & O_2 \end{array}$$

The interference of irrelevant variables for experimental group and controlling group should be avoided, such as the test, concurrent event, maturity, measuring tools, absence of examiners, and variance selection. Because the classrooms are not identical in the quasi experimental design for the controlling group, the controllable variables should ensure same instructor, same class duration, same curriculum materials, and same pre-test/post-test tools. Both groups have no absence of examiners and have accounting as a mandatory course (Borg & Gall, 1989; Campbell & Stanley, 1963). Because the pre-test result may not be identical and interferes the experimental result, the data obtained were analyzed by one-way analysis of covariance using the pre-test scores as covariate, the post-test scores as dependent variable, and the classes as independent variable.

4 Instrument Design and Execution Procedure

4.1 Design of Experimental Curriculum Materials

The experimental instrument used in this study is the teaching blog platform, named "the secret paradise of class 2A in the accounting information department." This blog, which was requested by the researcher and provided by PIXNET, was used for cooperative e-learning. The visitors' IP addresses recorded by the blog can be used for evaluating the e-files. Figure 1 shows the additional little circle function (the website address of class 2A of accounting information department is http://lmr0306.club.pixnet.net/). Through this blog,students can post their questions, interact with instructors and classmates, take a quiz, or submit their.

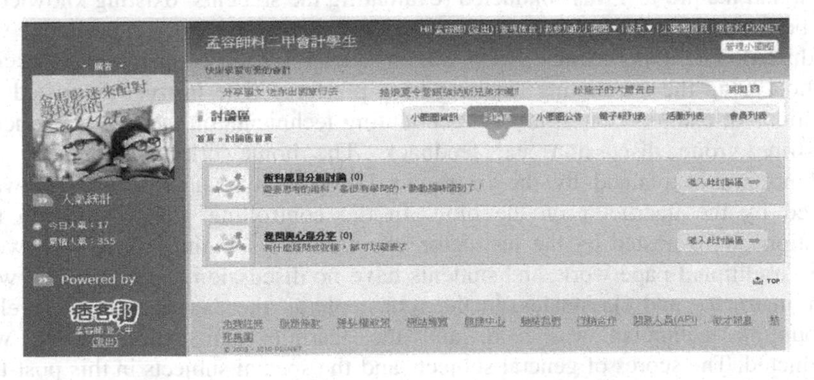

Fig. 1 The little circle of class 2A of accounting information department

4.2 Design of Testing and Measuring Survey

(1) Questions for test learning performance: There are total of 851 questions from 10 areas in the general subject question database of examination of level C accounting technician, which was provided by the Bureau of Employment and Vocational Training, Council of Labor Affairs, Executive Yuan; out of which, 80 questions were selected for pre-test. The test subjects include accounting textbooks of volume 1 & 2. Before the pre-test, the exam questions were analyzed in terms of knowledge, comprehension, application, and analysis using two way detailed analyses. The 45 students from the data processing classes of the 12th grade in a private vocational high school participated in the pre-test. The guideline for selecting pre-test questions was based on the evaluation standard proposed by the American examination researcher Ebel (Guo, 1989). The difficulty index was 0.2 to 0.8, and the discrimination index was over 0.25. Based on the analysis result using KR20 analysis designed by Tai (1992), 25 questions were extracted as the formal test questions, with Kuder-Richardson reliability coefficient of 0.85.
(2) Learning satisfaction survey: This study used the satisfaction survey created by Hsu (2008) to evaluate the learning satisfaction of the accounting class in the vocational high school. The measurement form has four parts: cooperative e-learning method, use of a teaching blog, learning interest of each accounting item, and overall learning satisfaction. The 5 point Likert scale was used in the survey. More points on the scale mean higher satisfaction. The value of Cronbach's α for each dimension in the pre-test survey was between 0.867 and 0.871. The average value of α was 0.871.

4.3 Experiment Execution Procedure

There were four phases in the teaching experiment. During the first phase, preparation tasks need to be completed, and four classes were given every week to both groups (e.g. teaching how to apply a personal account, practice on the blog functions, and group experiment groups). During the second phase, classes were given and the pre-test was conducted (evaluating the students' existing knowledge for both groups). During the third phase, the formal teaching experiment was conducted. The experimental group was instructed by the cooperative e-learning method using the accounting teaching blog platform. The instructors posted the questions of examination of level C accounting technician on the blog, conducted real-time group discussion and feedback. The homework submitted by the students was explained by the instructor immediately, and the answers were posted by the instructor on the blog. In the controlling group, although the questions were posted by the instructor, the students submitted their homework using traditional paperwork, and students have no discussion and interaction with their instructor and classmates. In the fourth phase, the examination of level C accounting technician was held, and the learning satisfaction survey was conducted. The scores of general subjects and the special subjects in this post-test were used to evaluate the learning achievement.

5 Data Analysis

5.1 Verify whether using the teaching blog platform for the students' cooperative e-learning is beneficial to their accounting skill test result.

(1) Pre-test：Table 1 shows that in the pre-test, the experimental group scored the mean of 58.24 while controlling group scored the mean of 54.17, with t=.991, p=.327, which did not reach a significant level, indicating that these two groups have the same level of accounting knowledge.
(2) Post-test：Table 1 shows that in the post-test, the experimental group scored the mean of 80.96 while controlling group scored the mean of 74.83, with t=3.118, p=.003, which reached a significant level, indicating that using cooperative e-learning through a blog is more helpful for improving students' performance than using the blog alone.

Table 1 Independent samples t-test analysis for the pre-test and post-test scores in the general subject

Group		Number of people	Mean	Standard Deviation	*t* value
Pre-test	Experimental group	25	58.24	12.387	.991
	Controlling group	24	54.17	16.215	
Post-test	Experimental Group	25	80.96	5.905	3.118*
	Controlling group	24	74.83	7.761	

*p<0.05 **p < .01 ***p < .001.

(3) Covariance analysis：Because the pre-test scores of these two classes were not identical, and it may interfere the experimental result, the data obtained were analyzed by one-way analysis of covariance using the pre-test scores as covariate, the post-test scores as dependent variable, and the classes as independent variable. By excluding the interference of the pre-test scores, the actual deviation of these two classes' post-test scores can be obtained. Table 2 shows that after excluding the interference of covariance, the difference of the independent variable (the post-test scores of the experimental group and the controlling group), F=8.960, p=.004, reached a significant level. This approves that using teaching blogs for cooperative e-learning is helpful for improving students' accounting skill test scores. Therefore, the empirical data supported hypothesis H1.

Table 2 Summary table of the covariance analysis on scores in the general subject with different teaching methods.

Source of variation	*SS*	*df*	*MS*	*F* value
Between-group	311.437	1	311.437	8.960*
Within-group	1,598.863	46	34.758	
Both groups	1,910.300	47		

*$p<0.05$ **$p < .01$ ***$p < .001$.

5.2. Students with cooperative e-learning method through the accounting teaching blog platform have significantly higher positive satisfaction than negative satisfaction.

To conduct the learning satisfaction survey, the questionnaire of "satisfaction scale of using a teaching blog for students' cooperative e-learning" was used after the learning performance was measured. The questionnaire has four parts: "cooperative e-learning method," "use of a teaching blog," "learning interest of each accounting item," and "overall learning satisfaction." Table 3 shows the analysis of the satisfaction level for each factor.

Table 3 Percentage of satisfaction for each factor

Factor	**Extremely agree**	**Agree**	**Neutral**	**Disagree**	**Extremely disagree**
Cooperativee-learning method	**10%**	**72.6%**	**18%**	**0%**	**0%**
Use of a teaching blog	**7.33%**	**74.67%**	**16.5%**	**0%**	**0%**
Learning interest of each accounting item	**16.5%**	**72.36%**	**15.64%**	**0%**	**0%**
Overalllearning satisfaction	**23.6%**	**68.4%**	**8%**	**0%**	**0%**

6 Conclusions and Suggestions

6.1. Using cooperative e-learning method through a teaching blog platform achieves significantly higher marks in the examinations of accounting technician than using the teaching blog alone.

The empirical results of this study shows that using cooperative e-learning method through a teaching blog platform achieves higher marks in the examinations of accounting technician. The international researchers, Chuang & Shen (2008), have investigated the learning achievement and learning satisfaction of three online classroom environment: traditional classrooms, blogs with knowledge sharing (cooperative learning), and blogs without knowledge sharing. The empirical results show that using cooperative learning through blogs has higher learning achievement and higher learning satisfaction. The empirical studies of researchers in Taiwan, such as Chiu (2004), and Chen (2008), also suggested that cooperative e-learning through digital platforms can improve students' scores. A blog can be used to record teaching activities and students' daily activities. It also provides a channel for academic exchange. The school can provide students the best source of knowledge from the log files saved on the server. Instructors can create a teaching blog with the cooperative learning principle, post the syllabus on the blog, and build an information database for students to search information. The functions of the blog, such as submitting homework, database of exam questions, and sharing learning experiences or topics, allows students to share their opinions and give feedbacks. With students' support and extensive participation of cooperative learning through a blog, the learning performance is enhanced.

6.2. Students with cooperative e-learning method through the accounting teaching blog platform have significantly higher positive satisfaction than negative satisfaction.

The empirical results show that students with cooperative e-learning method through the teaching blog platform have high satisfaction, indicating that it is acceptable for most students to engage in cooperative e-learning activities. Students thought that a teaching blog platform for cooperative e-learning enables good group discussion, provides communication tools and homework space, and makes learning more fun. For instance, classmates in the same group can discuss together on the assigned topics and take the accounting sample tests from the exam database after school. Using the cooperative e-learning through a teaching blog, students can encourage and help each other, which help them to overcome learning obstacles, motivate them to study accounting, and be responsible for their learning performance. Students can submit the assigned homework by "feedbacks" without the limitation of time and space, and instructors can provide comments right away, which foster students' willingness and motivation of learning.

References

1. Arslan, R.S., Sahin-Kizil, A.: How Can the Use of Blog Software Facilitate the Writing Process of English Language Learners? Language Learning 23(3), 183–197 (2010)
2. Baturay, M.H., Bay, O.F.: The Effects of Problem-Based Learning on the Classroom Community Perceptions and Achievement of Web-Based Education Students. Computers & Education 55(1), 43–52 (2010)
3. Black, L.: Blogging clicks with educators: Online forums make assignments, ideas more accessible to students and parents. Knight Ridder Tribune Business New (2007)
4. Borg, W.R., Gall, M.D.: Educational research: An introduction. Longman, New York (1989)
5. Campbell, D.T., Stanley, J.C.: Experimental and quasi-experimental designs for research. R. McNally, Chicago (1963)

6. Cheng, K., Chen, Y.F.: Example of Applying Incooperative e-Learning to Accounting Teaching. Business Professional Education Quarterly Journal 108, 23–28 (2008)
7. Chiu, J.L.: Teaching Performance Using Cooperative e-Learning – Case Study of Software Application Class in A Vocational High School. Master Thesis; Master Program of Department of Business Education, National Changhua University of Education; Unpublished; Changhu (2004)
8. Conhaim, W.W.: Blogging: What is it? Link-Up 19, 3 (2002)
9. Gallagher, J.: As Y'all Know: Blog as Bridge. Teaching English in the Two-Year College 37(3), 286–294 (2010)
10. Gong, H., Yan, S.: Study on Supportive Technologies to Construct Rural Digital Learning Platform. In: International Symposium on Intelligent Information Technology Application, vol. 1, pp. 794–798 (2008)
11. Guo, S.Y.: Methodology of Psychology and Education, 10th edn. Essence Publishing Company, Taipei (1989)
12. Harder, Reichardt: Throw another blog on the wire: Libraries and the web-blogging phenomena. Feliciter 49(2), 85 (2003)
13. Cheng, Y.C.: The Impact of Cooperative e-Learning on High School's English Earning Performance – Case Study in A Vocational High School. Master Thesis; Graduate School of Information Management, Nanhua University; Unpublished; Jiayi (2003)
14. Jheng, S.S., Lai, Y.H.: The Effects on Reader Services of Applying Blogs in High School Libraries in Taiwan. In: International Association of School Librarianship. Selected Papers from the Annual Conference, pp. 1–11 (2007)
15. Liu, Z.M.: Factors of Using Teaching Blogs By Middle School's Teachers. Master Thesis; Master program of Department of Industrial Education and Technology, National Changhua University of Education; Unpublished Thesis; Changhu (2009)
16. Tse, H.H., Chang, K.E., Sung, Y.T.: Using Blogs as a Professional Development Tool for Teachers: Analysis of Interaction Behavioral Patterns. Interactive Learning Environments 17(4), 325–340 (2009)
17. Oikonomidoy, E.: Conceptual Collective Online Reflection in Multicultural Education Classes. Multicultural Education & Technology Journal 3(2), 130–143 (2009)
18. Philip, R., Nicholls, J.: Group Blogs: Documenting Collaborative Drama Processes. Australasian Journal of Educational Technology 25(5), 683–699 (2009)
19. Sun, C.T., Lin, S.R.: Cooperative e-Learning. Psychology Publishing Company, Taipei City (2007)

A Study on Teaching Problems and Coping Strategies of Social Study Field Teachers

Chin-Wen Liao, Chih-Hao Chen, Li-Chu Tien, and Fang-Pin Lai

Department of Industrial Education and Technology,
National Chunghua University of Education. No.2, Shi-da road, Chunghua City, 500, Taiwan, R.O.C.
tcwliao@cc.ncue.edu.tw, chenuj22@hotmail.com,
m9316621@yahoo.com.tw, lion0829@gmail.com

Abstract. The purpose of this study is to explore the teaching problems that encountered by the junior high school teachers in social study field and the coping strategies based on individual background and school environment variables. Questionnaire survey method was established after thoroughly literature reviewing. A total of 264 valid questionnaires were returned while 342 questionnaires have been sent, and valid return ratio was 77.2%. The collected data was analyzed by statistical methods of frequency distribution, percentage, mean, standard deviation, T test, one-way ANOVA, Scheffe posteriori comparison and Pearson correlation. Conclusions and recommendations were made in accordance with the results of the analysis.Results were summarized as follows:1."Classroom Management" and "Teacher's Professional Competence" are viewed as the highly disturbing problems to junior high school teachers in social study field; and "positive" is their responding strategy when encountering problems.2.Teaching problems of junior high school teachers in social study field have significant differences due to different ages, positions, school located areas and school size.3.As junior high school teachers in social study field, Adjunct homeroom teachers are more actively seeking for solutions than adjunct administrative teachers; and adjunct administrative teachers are more actively looking for help than subject teachers.

Keywords: Social study field teachers, teaching problems, coping strategies.

1 Introduction

Following the guideline of Grade 1-9 Curriculum, "social study field" underwent a big change. The number of classes was reduced to half with the integrated curriculum consisting of geography, history, and civics. For a long time, these three junior high school subjects have been taught separately with different courses by teachers from different professional backgrounds, and each subject had a clear objective. The new curriculum requires an integrated teaching mechanism and makes it a lot more difficult for teachers to design the materials for teaching social study field.

L. Uden et al. (Eds.): Workshop on LTEC 2012, AISC 173, pp. 121–129.
springerlink.com

Since the Grade 1-9 Curriculum's implementation, the junior high school teachers of social study field have encountered more changes and challenges. Because the teaching hours of social study field are reduced while the textbook content is continuously expending and increasingly difficult, and the teaching software/hardware is not sufficient, the curriculum is facing the issue of marginality. These problems and pressures that teachers were facing should be investigated systematically to examine the actual teaching situation of social study field in junior high schools and provide these teachers coping strategies and practical teaching methods.

Prior studies also pointed out that the Grade 1-9 Curriculum created some challenges for teachers' professional mentality and competence. In the study of "Evaluating The Trial Condition And Result of Grade 1-9 Curriculum of Compulsory Education," Ya-Wen Chang (2001) pointed out that the schools in the pilot program expressed the difficulties of implementing the Grade 1-9 Curriculum, such as teachers' professional mentality and competence, related administration procedures, curriculum development, integration of human resources, training arrangement and participation, and additional burden for schools. In the case study of the difficulties of teaching social study field in Taitung county, Mei-Huan Lin (2008) stated that the implementation of the Grade 1-9 Curriculum caused the difficulties of teaching social study field, such as "teaching resources and equipments," "academic knowledge," "training opportunities," and "obtaining assistance from parents and communities", of which "teaching resources and equipments" was the most difficult factor. The Grade 1-9 Curriculum emphasizes the integrated teaching approach. However, the junior high school teachers who taught the social study field were originally trained only for a specific subject in the system that emphasized expertise and specialized field. To handle the integrated curriculum, teachers will face the challenges of professional competence, long teaching hours and too many classes in a week, work overload, and team teaching. Under the intense pressure that students are facing, questions arise about whether the teachers who teach the social study field have enough independence and specialty, and whether teachers with different background face different level of difficulties. Therefore, today when the education revolution has been implemented for ten years, it is necessary to develop coping strategies for teaching problems and challenges that teachers faced due to the Grade 1-9 Curriculum.

2 Research Design and Implementation

Based on the conclusions and results of the related theories and literatures by national and international researchers, the design of this study was proposed, and the empirical data was collected and analyzed statistically as follows:

2.1 Scope of this study: The 3 target areas in this study are Taichung City, Changhua County, and Nantou County.

2.2 Structure of this study: There are independent variables and dependent variables in this study.

2.2.1 The independent variables are: (A) teachers' background: gender, age, current position; (B) school's environmental background: school's location, school's scale.

2.2.2 The dependent variables are: (A) six aspects of teaching difficulties: teachers' professional competence, teaching practical experiences, teaching resources and equipments, textbook contents, classroom management, administrative support from school; (B) two coping strategies: asking for help, seeking solutions..

2.3 Samples of this study: The target group in this study was the Taiwanese teachers who taught social study field for the semesters in year 2010. The samples are the teachers who taught social study field at public schools in Taichung City, Changhu County, and Nantou County.

2.4 Tools used in this study: Survey was used in this study. Based on the literatures and the structure of this study, we referenced many scholars' questionnaire manuscripts (Rung-Yu Lin, 2010; Mei-Huan Lin, 2008) and compiled them into the "Questionnaire of Teaching Problems and Coping Strategies for Social Study Field Teachers in Junior High Schools."

2.4.1 Survey items: After the questionnaire draft was completed, the proposed "questionnaire for consulting experts" was revised by the instructing professor. The counselor in the social field was invited to assist in revising and giving opinions regarding to the content and the structure of the questionnaire draft. The valuable comments from these experts and scholars provided a good validation for the tools used in this study. The revised version was used for the test questionnaire, which was validated by experts. A multilevel random sampling survey was conducted in this study. Each level is proportional to the number of teachers who taught social study field in the junior high schools in Taichung, Changhua, and Nantou. The sample teachers were then randomly selected. Out of the 156 questionnaires distributed, 144 copies were returned. After discarding the invalid questionnaires and those with incomplete answers, a total of 141 valid questionnaires were received, which consisted of 90.4% of the test questionnaires.

2.4.2 Factor analysis: The "principal component analysis" was used in this study to anylyze the factors. The orthogonal rotations was performed using varimax method to analyze the common factors' components. The factors with eigenvalue larger than 1 were selected. Each questionnarie item was categorized into different aspects, the name of each aspect was specified. For analyze the factors of "teaching problems," the KMO value was .786, and 6 factors were selected with the accumulated explanatory variance of 61.356%. To analyzing the factors of "coping behaviors", the KMO value was .671, and 2 factors were selected with the accumulated explanatory variance of 62.759%.

2.4.3 Reliability: Regarding to "teaching problems," the α coeffeicient of overal scale was .893; the inner consistency coeffients of the six sub-scales were .806, .707, .871, .714, .705, and .856 respectively. Regarding to "coping behavors," the α coeffeicient of overal scale was .761; the inner consistency coeffients of the 2 sub-scales were .598 and .739 respectively. The Cronbach's α coeffient of overal scale was .973, which indicates great inner consistency.

3 Data Analysis and Results

The survey target in this study was the social study field teachers in 142 public junior high schools in Taichong, Changhua, and Nantou. 342 questionnairs were distributed on April 6, 2011. By May 12th, 282 questionnaires were received with the returned rate of 82.5%. After discarding 18 invalid questionnaires, there were 264 valid questionnaires, which is showing in table 1.

Table 1 Questionnaire distribution and collection

City/County	Class Number	Class Raitio	Sampling Questionnaires	Returned Quesionnaries	Valid Questionnaires
Taichong City	3528	61%	186	151	140
Changhua County	1540	27%	111	92	89
Nantou County	701	12%	45	39	35
Total	5769	100%	342	282	264

3.1 Teaching problems and coping strategies of junior high school teachers of social study field

3.1.1 Teaching problems of junior high school teachers of social study field: The junior high school teachers of social study field perceived more difficulties on "classroom management" and "teachers' professional competence," which were on the scale of "highly difficult." The mean score was 3.14, which was on the scale of "somewhat difficult."

3.1.2 Coping strategies of junior high school teachers of social study field: The two coping strategies proposed in this study are "asking for help" and "seeking solutions," which scored mean of 3.24 and 2.88 respectively. The overall scale scored mean of 3.14, indicating that junior high school teachers of social study field were actively coping with the problems. To cope with the difficulties, the teachers tended to "asking for help" more than "seeking solutions," as shown in table 2.

Table 2 Summary of teaching problems and coping strategies of junior high school teachers of social study field

	Aspect	Mean score of each question	Ranking	Degree
Teaching Problems	Teachers' professional competence	3.71	2	Highly difficult
	Teaching practical experiences	2.84	6	Somewhat difficult
	Teaching resources and equipments	3.11	3	Somewhat difficult
	Textbook contents	2.98	4	Somewhat difficult
	Classroom management	3.89	1	Highly difficult
	Administrative support from school	2.92	5	Somewhat difficult
	Overall teaching problems	3.14		Somewhat difficult
Coping Strategies	Seeking solutions	2.88	2	Actively
	Asking for help	3.24	1	Actively
	Overall coping strategies	3.14		Actively

N=264

3.2 The relation between teachers' background and teacher perceived teaching difficulty

The variance analysis of teacher perceived teaching difficulties for different personal background is showing in table 3: (A) the statistical analysis of teacher' gender and teaching problem: no significant difference exists in different genders for junior high school teachers of social study field; (B) the statistical analysis of teacher' age and teaching problem: in terms of "teachers' professional competence," teachers who age between 31 to 40 perceived higher difficulties than those age 30 years old or younger; (C) the statistical analysis of teachers' position and teaching problems: in terms of "teachers' professional competence," homeroom teachers perceived higher difficulties than teachers with administrative responsibilities and other full-time teachers.

Table 3 Analysis of teaching problems and personal background of junior high school teachers of social study field

Aspect	Gender	Age	Teachers' position
Teachers' professional competence		(2)>(1)	(2)>(1),(3)
Teaching practical experiences			
Teaching resources and equipments			
Textbook contents			
Classroom management			(3)>(1)
Administrative support from school			(2),(3)>(1)
Overall teaching problems		(2)>(1)	(2),(3)>(1)

Note: N=264; $p<.05$; teachers' age groups: (1) 30 years old or younger, (2) 31 to 40 years old, (3) 41 to 50 years old, (4) 51 years old or older; teachers' current position: (1) teachers with administrative responsibilities, (2) homeroom teachers, (3) full-time teachers.

3.3 The relation between teachers' background and coping strategies

The variance analysis of the relation between strategies of coping teaching problems and teachers' backgrounds is showing in table 4: regarding to the strategies of coping teaching problems faced by junior high school teachers of social study field, both of the two aspects, "seeking solutions" and "asking for help," reached a significant level. The homeroom teachers were more actively "seeking solutions" than the teachers with administrative responsibilities. The teachers with administrative responsibilities were more actively "asking for help" than other full-time teachers.

Table 4 Variance analysis of the relation between strategies of coping teaching problems and personal background of junior high school teachers of social study field

Aspect	Gender	Age	Position
Seeking solutions			(2)>(1)
Asking for help			(1)>(3)
Overall coping strategies			

Note: N=264; $p<.05$; teachers' current position: (1) teachers with administrative responsibilities, (2) homeroom teachers, (3) full-time teachers.

3.4 The relation between school environment and teaching problems

3.4.1 The statistical analysis of the relation between school locations and teaching problems: the difference of teaching difficulties, in terms of "teaching resources and equipments," "administrative support from schools," and "overall teaching problems," for different school locations reached a significant level. This indicates that teachers in schools located in general areas perceived significantly higher difficulties than those in schools located in remote areas.

3.4.2 The statistical analysis of the relation between school size and teaching problems: the difference of teaching difficulties perceived by junior high school teachers of social study field, in terms of "teachers' professional competence," "teaching resources and equipments," "classroom management," "administrative support from schools," and "overall teaching problems," for different school size reached a significant level.

Table 5 Variance analysis of the relation between difficulties in teaching social study field and environmental backgrounds of junior high schools

Aspect	School location	School size
Teachers' professional competence		(2), (3), (4)>(1) (3)>(2)
Teaching practical experiences		
Teaching resources and equipments	(1)>(2)	(3)>(2),(4)
Textbook contents		
Classroom management		(3), (4)>(1)
Administrative support from school	(1)>(2)	(3), (4)>(1)
Overall teaching problems	(1)>(2)	(3), (4)>(1)

Note: N=264; School location: (1) general area, (2) remote area; School size: (1) 17 classes or less, (2) 18 to 44 classes, (3) 45 to 61 classes, (4) 62 classes or more.

4 Conclusions and Suggestions

4.1 Conclusions

(1) The teaching difficulties of junior high school teachers of social study field scaled as "middle high": The six aspects of teaching problems of junior high school teachers of social study field are "classroom management," "teachers' professional competence," "teaching resources and equipments," "textbook contents," "administrative support from schools," and "practical teaching experiences," in the order of the highest to the lowest. All aspects and "overall teaching problems" scaled as "middle high."

(2) When facing teaching problems, junior high school teachers of social study field will actively seek solutions and ask for helps: The active coping strategies suggested in this study are "asking for help" and "seeking solutions." "Asking for help" scaled higher than "seeking solutions," and both scaled as "actively."

(3) Teachers with different age, and position perceived significantly different level of teaching difficulties: (A) Those who age 31 to 40 years old perceived higher difficulty in terms of "teachers' professional competence" and "overall teaching problems" than those age 30 years old or younger. (B) In terms of "teachers' professional competence," homeroom teachers perceived higher difficulty than those with administrative responsibilities. In terms of "classroom management," full time teachers perceived higher difficulty than those with administrative responsibilities. In terms of "administrative support from schools," homeroom teachers and other full-time teachers perceived higher difficulty than those with administrative responsibilities.

(4) In schools with different location, and size, teachers perceived significantly different level of teaching difficulty: (A) Teachers in schools located in general areas perceived higher difficulty than those in remote areas, in terms of "teaching resources and equipments," "administrative support from schools," and overall teaching problems. (B) In terms of "teachers' professional competence," teachers in medium size schools (18 to 61 classes) and large schools (62 classes or more) perceived significantly less difficulty than those in small schools (17 classes or less); teachers in medium-to-large schools (45 to 61 classes) perceived significantly higher difficulty than those in small-to-medium schools (18 to 44 classes), with both scaled as "highly difficult." (C) In terms of "teaching resources," teachers in medium-to-large schools perceived higher difficulty than those in small-to-medium schools, with both scaled as "highly difficult." (D) In terms of "classroom management," teachers in medium-to-large schools perceived higher difficulty, which scaled as "highly difficult," than those in small and large schools. (E) In terms of "administrative support from schools" and overall teaching problems, teachers in medium-to-large schools perceived higher difficulty, which scaled as "somewhat difficult," than those in small and large schools.

(5) Teachers with different positions have different coping strategies for teaching social study field in junior high schools: homeroom teachers are more actively "seeking solutions" than teachers with administrative works; teachers with administrative works are more actively "asking for help" than other full-time teachers.

4.2 Suggestions

(1) Suggestions to school administrations: (A) Build an internal and inter-school teaching resource network: encourage teachers to create teaching information files, motivate teachers to share teaching materials, establish teaching resources,

and build a internal network platform to share resources in the school where teachers can use these high-tech equipments to discuss with each other, share resources and experiences, and effectively prepare the teaching materials. (B) Provide appropriate back support: The survey result shows that schools in general areas did not give teachers enough administrative support, which caused high perceived teaching difficulty. We suggest the administrative staff to pay more attention to social subject teachers' need and set up additional classroom for teaching social subject. The administrative staff should provide appropriate back support to teachers to enable teachers and students to conduct teaching activities in the most convenient and suitable environment.

(2) Suggestions to teachers: (A) Improve social study field teachers' ability in classroom management: the survey result shows that most teachers still felt that they do not have enough professional competence to manage a classroom. Therefore, schools and education authorities should focus on enhancing teachers' professional knowledge and skills for classroom management and actively provide workshops and trainings for teachers. (B) Make a good use of internet for supplemental teaching: the survey result shows that because teachers only had three courses in a week to teach social study field for each class, they can not give supplemental teaching for those students who were left behind or test students' learning performance. We suggest teachers to create a blog where they can upload the recorded teaching videos for student to review and catch up.

References

1. Billings, A.G., Mooszya, R.H.: Coping stress and social resources among adults with unipolar depression. Journal of Personality and Social Psychology 46(4), 877–891 (1994)
2. Fernando, D.B., Amparo, G.A.: Barriers Perceived by Teachers at Work, Coping Strategies, Self-efficacy and Burnout. The Spanish Journal of Psychology 13(2), 637–654 (2010) (Spanish)
3. Folkmane, S., Moskovita, J.T.: Coping: Pitfalls and promise. Annual Review of Psychology 55, 745–774 (2004)
4. Lin, M.-H.: Strategies and Supplement Measures of Coping Teaching Problems for Junior High School Teachers of Social Study Field in Taitung Country. Master Thesis; Master program of Cultural and Educational Administration, Department (Graduate School) of Education, National Taitung University; Unpublished; Taitung (2008)
5. Robbins, S.P.: Organizational behavior, 9th edn. Prentice Hall, N.J (2001)
6. Lin, R.-Y.: Teaching Problems and Coping Strategies for Chinese Teachers in Junior High Schools in Yilan. Master Thesis; In-Service Master Program; Graduate School of Education, National Dong Hwa University; Unpublished; Hualian (2010)
7. Wiles, J., Bondi, J.: Curriculum development:A guide to Practice, 6th edn. Prentice-Hall, New Jersey (2002); Chang, Y.-W.: Evaluating the Trial Condition and Result of Grade 1-9 Curriculum of Compulsory Education. Master Thesis; Department of Education, National Kaohsiung Normal University; Unpublished; Kaohsiung (2001)

The Impact of Integrating Information Technology into Teaching on Teacher Education of Taiwan's Secondary Education

Chin-Wen Liao, Sho-Yen Lin, Li-Chu Tien, and Yi-Chen Chang

Department of Industrial Education and Technology,
National Chunghua University of Education. No.2, Shi-da Road, Changhua City 500, Taiwan, R.O.C.
cwliao.robert@msa.hinet.net, {linshoyen,yd3174}@gmail.com, m9316621@yahoo.com.tw

Abstract. The purpose of this study was to investigate the impact of integrating information technology (IT) into teaching on teacher education of Taiwan's secondary education. The study conducted statistical analysis of questionnaires with the data collected from 356 pre-service teachers. It was indicated that a significant positive correlation was found between teaching ability and attitudes. Analysis and discussion were conducted on qualitative opinions provided by 12 pre-service teachers, 10 novice teachers, and 6 persons who dropped out of teacher education, which help further clarify the issue and teachers' competence as a teacher in the future.

Keywords: pre-service teacher, e-learning, integrating information technology (IT) into teaching, teacher training institutions.

1 Introduction

IT is one of the fastest growing technologies in Taiwan, and its development level is the indicator used to measure the modernization of the country. The impact of IT on modern human society is almost limitless and closely correlated to each person's daily life, such as the mobile internet, personal digital assistant (PDA), electronic commerce, online auction shopping, video conference, distance teaching, real-time multicast teaching, e-learning, online teaching platform, and so on, which prove that the rapid development of IT has generated enormous impact on human life and educational environment! (Ministry of Education, 2011; Alavi et al., 1998; Brownell & Brownell, 1991)

Teachers are the soul of education and the success of IT was determined by their teaching ability and attitude of IT integration. In recent years, education administrations have invested a lot of education funds and actively promoted IT education (Ministry of Education, 2011). Since the Teacher Education Law was promulgated in 1994, teacher education channels have been developed toward directions of opening and diversity, including teacher education centers and education credit courses in Normal Universities and in general universities.

L. Uden et al. (Eds.): Workshop on LTEC 2012, AISC 173, pp. 131–143.
springerlink.com

The number of unemployed teachers with teacher qualifications has far exceeded the demand for secondary school teachers. According to the statistic report (2010) by Ministry of Education, the average acceptance rate in the selection of secondary school teachers was about 1-4%, and even the IT capability was set as the admission threshold for the selection of teachers. In other words, the reserve teachers in teacher education centers taking IT education concepts, application of IT skills, integrating IT into teaching ability, as well as attitude toward IT integrated into teaching were all key factors to determine success or failure on the promotion of information education today (Soner, 2000; Moersch, 1995). And understanding of the pre-service teachers' ability and attitude for integrating IT into teaching would help spread information education around at the high school teaching field, and cultivate the citizens with modern IT literacy.

According to the above mentioned backgrounds and motivations, the purposes of this study were as follows:

1.1 To understand the current situation of pre-service teachers' ability and attitude for integrating IT into teaching.

1.2 To analyze the differences on abilities and attitudes of integrating IT into teaching for pre-service teachers of different background variables.

1.3 To explore the correlation of pre-service teachers' ability and attitude toward integrating IT into teaching.

1.4 To analyze the impact of integrating IT into teaching based on pre-service teachers' teaching abilities and attitudes.

2 Material and Methods

2.1 Research Framework

Based on the theoretical foundation established from the research purposes and literature review, and aimed at the pre-service teachers' teaching ability and attitude for integrating IT into teaching to conduct the questionnaire survey, the purpose was to understand the current situation and the differences among pre-service teachers from different personal backgrounds regarding their teaching abilities and attitudes on integrating IT into teaching, and to explore the correlation between the two as well.

The independent variables were the personal backgrounds, including gender, school type, School Department Category, e-learning, and information network experiences. The dependent variables were the abilities and attitudes of integrating IT into teaching, in which the ability for integrating IT into teaching including IT recognizing, understanding of Internet technology connotation, the search and evaluation of e-teaching resources, and the application and assessment of e-learning; The attitude for integrating IT into teaching including cognitive domain, affective domain, and psychomotor domain. The research framework was shown in Figure 1.

2.2 The Samples

2.2.1 The questionnaire survey was adopted to collect quantitative data, taking a total of 1,466 people from seniors of normal universities and the sophomores who selected Secondary Education Program in the teacher education center of the general public and private universities in Taiwan as the population. By proportionally stratified random sampling, 356 valid samples were collected.

2.2.2 The interview method was adopted to collect qualitative data, drawing 12 pre-service teachers, 10 novice teachers, and 6 persons who dropped out of teacher education as the samples. Based on quantitative statistical analysis, the related factors that affected pre-service teachers on integrating IT into teaching were further clarified.

2.3 The Research Tools

"questionnaire survey of the impact of integrating IT into teaching on pre-service teachers teaching abilities and attitudes" and "interview questionnaires of the related factors that affect pre-service teachers on integrating IT into teaching" were adopted to collect data that based on the results of exploring the related literatures to develop the research tools for this study, and amended by experts to be a pre-test questionnaire, and the implementation of the pre-test questionnaire after, reliability and validity were tested, and then a formal questionnaire was completed. The ability scale reliability of the questionnaire was .887(α coefficient), and the attitude scale reliability was .894 (α coefficient), as shown in Table 1.

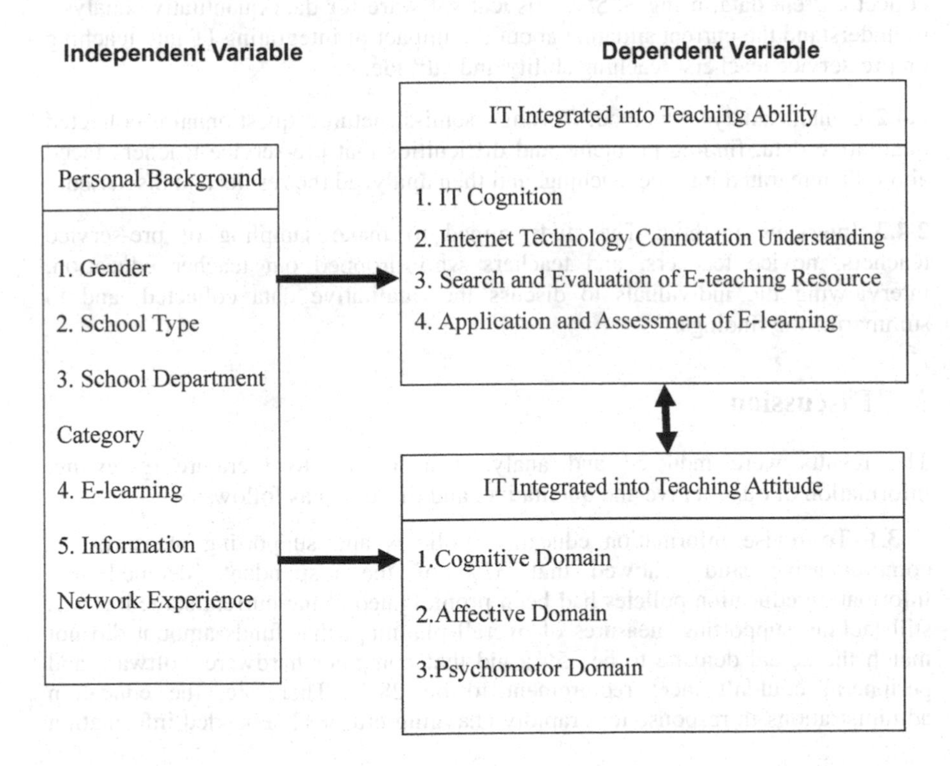

Fig. 1 The Research Framework

Table 1 Reliability Analysis Result of Integrating IT into Teaching Ability & Attitude

Integrating IT into Teaching Ability		Integrating IT into Teaching Attitude	
Dependent Variable	α	Dependent Variable	α
IT Recognizing	.740	Cognitive Domain	.829
Internet Technology Connotation Understanding	.754	Affective Domain	.703
Search and Evaluation of E-teaching Resource	.697	Psychomotor Domain	.896
Application and Assessment of E-learning	.832		
Full Scale of Ability	.887	Full Scale of Attitude	.894

2.4 Research Methods

2.4.1 Questionnaire survey method: The questionnaire survey was adopted to collect current data, using SPSS statistical software for data quantitative analysis to understand the current situation about the impact of integrating IT into teaching on pre-service teachers' teaching ability and attitude.

2.4.2 Content analysis method: Through semi-structured questionnaire collected qualitative data, finding problems and difficulties that pre-service teachers faced about IT integrated into the teaching, and then analyzed the results and discussed.

2.4.3 Interview method: The study aimed to make sampling of pre-service teachers, novice teachers, and teachers who dropped out teacher education, interviewing the individuals to discuss the qualitative data collected, and to summarize the findings.

3 Discussion

The results were inducted and analyzed according to literature reviewing, information of quantitative and qualitative, and discussed as follows:

3.1 To revise information education policies and supporting measures: A comprehensive study showed that 37% of the respondents deemed that information education policies had been promulgated at the current stage but was still lacking supporting measures of overall planning; that funds amount did not match the actual demand to be 35%; and that computer hardware, software and peripheral couldn't meet requirement to be 28%. Therefore, the education administrations in response to a rapidly changing era of IT amended information education policy and related supporting measures, aiming at the issues of integrating IT into teaching to establish an example and teaching materials for all disciplines firstly, observing the results through the teaching experiment, secondly to widen the Annual Plan funds to teacher education center for the secondary education to meet the promotion direction of education policy and actual teaching needs.

3.2 To train seeded teachers of integrating IT into teaching: The study results indicated that the information education policy focused more on updating of computer hardware and software equipments, but lacked a complete training mechanism for the seeded teachers of integrating IT into teaching, resulting in being ineffective on the promotion and applications of IT education. Therefore, education administrations should first train teachers to possess the concept of integrating IT into teaching related technologies and combine it with learning theory to enhance teachers' willingness to apply IT in the teaching field, giving them the task to continue promoting it, 65% of the respondents emphasized its importance.

3.3 To establish resource website for integrating IT into teaching: To integrate teaching websites and e-campus resources including the domestic and the international to build a good teaching software, diversity teaching materials, integrating teaching strategies, and to establish supporting community for integrating IT into teaching, advocating network cooperative learning, providing digital information, and teaching experience exchange platform, 68% of the respondents accented its importance.

3.4 To upgrade computer network equipment, reduce the e-learning gap: The data analysis showed that the e-learning gap was still widespread between high schools in Taiwan, especially in remote districts. Therefore, the educational administration should actively promote IT education, considering grant funds for purchasing or updating information software and hardware facilities, and enrich online education resources and information systems maintenance expenses, to provide computer network equipments for a sufficient supply of e-learning resources and effectively achieved the purpose of integrating IT into teaching.

3.5 To update network equipment, to improve the e-learning environment: The study results indicated that pre-service teachers who used computers weekly with more hours on the Internet had a better ability and a good attitude to integrate IT into teaching. Accordingly, teacher education center should improve the information network equipment and offer equipment using time and space to make pre-service teachers use the e-learning environment adequately.

3.6 To establish teaching interactive platform to facilitate the sharing of teaching demonstration: The teacher education center should hold integrating IT into teaching related learning or teaching demonstration activities to encourage pre-service teachers to participate. The learning activities most needed by 78% of pre-service teachers included: digital materials making, teaching web design, network resource management, teaching files, distance learning, teaching demonstration, and so on. Studying results should be effectively managed and maintained in the information network platform to enable teachers' mutual exchange and sharing.

3.7 To improve the quality of teaching resources, effectively apply e-learning effectiveness: The study results showed that the pre-service teachers have been participated in integrating IT into teaching lesson plan design, their ability of integrating IT into teaching would be better than those who have not involved in. Teacher education center should consecrate to enhance the quality and quantity of information network teaching resources and encourage pre-service teachers to effectively use these resources on lesson plans design and teaching activities, in order to get practice experience of integrating IT into teaching, which would help improve pre-service teachers' competence as a teacher.

3.8 To build remote video environment to develop professional growth strategies: The inductive analysis of qualitative data found that 81.5% of the respondents deemed that the teacher education center should build remote video systems to comply with the teaching facilities requirements in the era of information technology. Through research assessment and discussion, remote video should be developed for uses on the various disciplines teachers professional growth model for the establishment of partnership between universities and high schools to strengthen the combination of teachers teaching theory with practice.

3.9 To enrich the computer network related skills to enhance capacity of IT applications: The study results pointed out that pre-service teachers who used computers weekly with more hours on the Internet had a better ability and a good attitude to integrate IT into teaching. Accordingly, it was recommended that during the study period, pre-service teachers should enhance their ability to the related integrating IT into teaching.

3.10 To select integrating IT into teaching curriculum to enrich pedagogical content knowledge(PCK) in professional fields: The study results indicated that the pre-service teachers who attended courses of integrating IT into teaching and of the relevant information network had a better performance ability on PCK of their integrating IT into teaching. Therefore, pre-service teachers in particular should select more teacher education center offered courses in order to develop well the ability for integrating IT into teaching, and at the same time established the correct concept of integrating IT into teaching: taking students as the center, to enhance the effectiveness of teaching as objectives.

3.11 To participate in education trainings to enrich IT knowledge and skills: The study found that pre-service teachers would have a better ability of integrating IT into teaching if they have participated in information networks training courses. Hence, pre-service teachers should have a more positive attitude towards learning, to enrich the IT ability, enhancing their professional development, and possessed information network technology application capabilities to undertake the post as a secondary school teacher, thus achieved the purposes on teaching benefits for teachers as well as students.

3.12 To establish teaching portfolio, integrating IT into teaching: The data analysis showed that pre-service teachers who have begun to use teaching portfolio had a better ability for integrating IT into teaching. They could take opportunities of assignment delivery and presentation, or the teaching demonstration, and educational practices, using integrating IT into teaching to establish their own unique teaching design model, or collecting the paradigm database that was established by master teachers, that could provide pre-teachers valuable teaching experiences for reference when they faced the issue of teaching.

3.13 To shape cooperative learning community and sharing e-learning experiences: According to the results of this study, pre-service teachers were willing to engage in cooperative learning, collaborative teaching and sharing regarding integrating IT into teaching experiences. During the period of teacher education, a teaching research group or a virtual network-learning community was established to provide a cooperative learning exchange platform to promote mutual discussion and sharing of experiences and creativity on integrating IT into teaching.

3.14 To integrate IT into teaching to stand out from the teacher selection: 98% of the novice teachers believed that integrating IT into teaching has been the basic capacity what secondary school teachers had to have, including the digital teaching materials making, teaching web design, creating teaching portfolios, applying network teaching platform, managing the network resources, and so on. These capabilities were not only the threshold for pre-service teachers to be the formal teachers, but were also the capabilities in response to the impact of future technological changes on the teaching environment, which would drive teachers to learn the aforementioned IT abilities constantly.

3.15 To participate in learning activities to enhance teacher's PCK: The study results indicated that pre-service teachers have learned more information related courses, their abilities on integrating IT into teaching were better. In addition, the literature review results pointed out that with use experiences of the computer and Internet would enhance teachers' ability, willingness, and attitude to use IT. Therefore, teachers had to cultivate the spirits of active learning and energetic participation to seize every opportunity to study IT.

3.16 To organize the teaching observation groups to promote the exchange and sharing of teaching experiences: The study found that pre-service teachers would have a better ability on integrating IT into teaching if they have ever observed the teaching demonstration; secondary school teachers could share their teaching experiences, discussing mutual experiences of integrating IT into teaching by observation of teaching demonstration among colleagues, and even organized observation teams of integrating IT into teaching, which would be of great promotion on integrating IT into teaching.

4 Results and Conclusions

According to the purposes, this study was by means of statistical analysis of data collected through questionnaire surveys obtaining the results as below:

4.1 Basic Data Analysis

Aiming at valid samples analyzed 5 basic data: gender, school type, School Department Category, e-learning experiences, and information network experience; in which e-learning experience was further divided into five items:

credits in attending courses related to integrating IT into teaching, credits in attending courses related to information networks, studying experience of participating in information network related courses, participating in the lesson plans design of integrating IT into teaching, and observation of integrating IT into teaching demonstration. Information network experience was divided into two items: computer hours and Internet time weekly, which were as shown in Table 2.

Table 2 Independent Variable of Valid Samples for Distribution List

Ind. Variable	Item	Sample	Percentage
Gender	Male	124	40.8%
	Female	180	59.2%
School Type	Normal University	175	57.6%
	University in General Education Program	129	42.4%
School Department Category	College of Art	239	78.6%
	College of Science	65	21.4%
IT Integrated into Teaching Related Course	0 Credit	52	17.1%
	1 -4 Credit	171	56.3%
	5 -8 Credit	51	16.8%
	9 Credit up	30	9.9%
Information Network Related Course	0 Credit	58	19.1%
	1 -4 Credit	177	58.2%
	5 -8 Credit	42	13.8%
	9 Credit up	27	8.9%
Information Network Related Course Learning Experience	Yes	59	19.4%
	No	245	80.6%
Participating in IT Integrated into Teaching Lesson Plans Design	Yes	95	31.3%
	No	209	68.8%
Observing IT Integrated into Teaching Demonstration	Yes	122	40.1%
	No	182	59.9%
Computer Hour and Internet Time Weekly	0 - 7 Hr	45	14.8%
	8 -14 Hr	55	18.1%
	15 -21 Hr	52	17.1%
	22 -28 Hr	65	21.4%
	29 -35 Hr	36	11.8%
	36 Hr up	51	16.8%
Accumulative Computer Use Time	within 4 Yr	12	3.9%
	5 - 7 Yr	60	19.7%
	8 -10 Yr	113	37.2%
	11 Yr up	119	39.1%

N=304

This study focused on the pre-service teachers, the current situation, the difference, and the relationship regarding their ability and attitude on IT integrated into teaching. The achieved results were shown in Table 3 and described below.

Table 3 Purposes and Results Contrast List

Purpose	Result
1.understanding the current situation of pre-service teachers' ability and attitude on IT integrated into teaching	1-1ability reaching above the average level (M=3.70) 1-2attitude up to affirmative and positive level (M=3.63)
2.discussing the difference of pre-service teachers' ability on IT integrated into teaching with the different personal background variable	2-1 school type and school department category without significant difference 2-2 male teaching ability better than female's 2-3 the more study credits, the more teaching abilities 2-4 joining in information network learning, participating in teaching lesson plans design, observing teaching demonstration, and so on, having a better teaching ability 2-5 the more computer hours and Internet time, the more teaching ability
3.discussing the difference of pre-service teachers' attitude on IT integrated into teaching with the different personal background variable	3-1 being significant difference items including gender, school type, school department category, joining in information network related courses studying, participating in IT integrated into teaching lesson plans design, observing IT integrated into teaching demonstration, computer hours and Internet time 3-2 the more related IT integrated into teaching credits and computer network related credits, the more positive significant difference 3-3 computer hours reaching significant difference, computer time and attitude being positive correlation

Table 3 (*continued*)

4.discussing the relationship between pre-service teachers' ability and attitude on IT integrated into teaching	4-1significant correlation between the two, its Pearson product-moment correlation coefficient γ=.643, significance level p=.000, representing positive correlation
5.inductively analyzing the impact to pre-service teachers' teaching ability and attitude on IT integrated into teaching	5-1 IT integrated into teaching influence pre-service teachers' CK, PK, and PCK in their professional field. 5-2 Using pedagogic theory in teaching field will influence the performance of teaching ability and attitude, including teachers' teaching change progress, teachers' belief, teachers' emerging technology PCK. 5-3 Valuing the impact of IT integrated into teaching to pre-service teachers or pre-service enterprise talent cultivation, to meet IT society development trend.

Taiwan pre-service teachers' ability of integrating IT into teaching achieved above the average level, and the attitude was affirmative and positive. The statistical analysis results were described as follows:

4.2 Pre-service teachers' ability on integrating IT into teaching achieved above the average level (M=3.70)

4.2.1 "understand Internet technology connotation" got the highest scores (M= 4.11), in which to "use network technology to search for the pedagogy related supporting teaching materials" and "share network technology experience with others to promote cooperative learning opportunities" were the best in ability; but the ability to "set up message boards or discussion areas for students, teachers, and parents to interact" got the lowest scores (M=3.92), just the better level.

4.2.2 Next was "the search and assessment of e-teaching resources" (M=3.72), in which the ability of "choose appropriate teaching resources network for students," was the better, the ability to "select proper e-learning website to collect pedagogic software that is appropriate for the individual grade and classes" was weaker.

4.2.3 The third was "the application and assessment of e-learning" (M=3.55), in which the ability to "use word processing software to edit the required teaching activity design, teaching sheets, and other documents" and "use power point software, such as making PPT files and oral presentation," were the best and respectively in the leading position of the full scale; but the ability to "use website editing software to make pedagogical website" was ranking on the end of the scale.

4.2.4 The final was "IT recognizing" (M=3.42). The ability was above "normal" level, in which the ability to "understand the relative norms in use of information network" was good; and the worst was "Understand the Ministry of Education implemented information basic literacy indicators for secondary school teachers" with the average 3.07. The ability was of "normal" level, the lowest in the full scale.

4.3 The attitude toward integrating IT into teaching: "full scale of attitude" achieved affirmative and positive level (M=3.63). The sequence of attitude scales was described as follows:

4.3.1 The best was the "cognitive domain," in which the "implementation of integrating IT into teaching can provide for a more diversity assessment" was the best; and the average of "the implementation of integrating IT into teaching will increase the burden on teachers" (reverse item) was 3.87, reaching "normal" level or above.

4.3.2 Next was "affective domain," in which the item "I'm afraid to use computer-related equipment" (reverse item) was the best (M=2.48); as for the item "being confident of the implementation of integrating IT into teaching," the average was 3.46, indicating a positive attitude.

4.3.3 The last was "psychomotor domain." (M=3.39), reaching the above medium level, in which "often participated in the implementation of integrating IT into teaching" was the best; but "often participated in the study activities of integrating IT into teaching" was to be further strengthened.

4.4 The differences in the ability of integrating IT into teaching among pre-teachers from different background variables

4.4.1 As for the ability in response to integrating IT into teaching, under the variable of background, the ones with better performances were male, had more credits in the courses of regarding integrating IT into teaching and computer network, participating in information network related courses, and joined in integrating IT into teaching lesson plans design, observing the demonstration of integrating IT into teaching, and had spent more hours in using computers and Internet time weekly.

4.4.2 As for "IT recognizing," with the outstanding ability in performance were those who had more credits in the courses of regarding integrating IT into teaching, and joined in integrating IT into teaching lesson plans design, observing the demonstration of integrating IT into teaching, and had spent more hours in using computers and Internet time weekly.

4.4.3 As for "understand the connotation of network technology," the ones with the excellent ability in performance were graduates from a Normal University, and had more credits in the courses of computer network, and joined in integrating IT into teaching lesson plans design, observing the demonstration of integrating IT into teaching, and had spent more time in using computers.

4.5 The differences in pre-service teachers' attitude toward integrating IT into teaching under the variable of different background

4.5.1 As for the attitude in response to integrating IT into teaching, the ones with better performances were those who had more credits in the courses related to the computer network and integrating IT into teaching, and had spent more time in using computers.

4.5.2 As for the "cognitive domain" attitude, the ones with a better attitude in response were those who had observed the demonstration of integrating IT into teaching and had more time in using computers.

In conclusion, the teacher education center during the period of training pre-service teachers should emphasize guidance strategies of teaching practice to promote pre-service teachers to link practice and theory. Summarizing results of this study, that IT integrated into teaching affected pre-service teachers' CK, PK, and PCK in their professional field (Dexter, Anderson & Becker, 1999; Tamir, 1991). Especially, teachers applied the pedagogical theory in teaching field would influence their performance of teaching ability and attitudes, including teachers' teaching change progress, teachers' belief, and teachers' emerging technology PCK (Ertmer, Addison, Lane, Ross & Woods, 1999). Therefore, no matter they were education administrations, teacher education center, pre-service teachers, novice teachers, or those who dropped out of teacher education, all should pay attention to the impact of integrating IT into teaching on pre-service teachers or enterprise talent cultivation, so could meet the development and trend of IT society.

Acknowledgements. This work has been partially supported by National Science Council, Executive Yuan, Taiwan.

References

1. Alavi, M.B., Borkowski, E.Y., Norman, K.: Emergent patterns of technology/ learning in electronic classroom. Educational Technology Research and Development 46(4), 23–42 (1998)
2. Brownell, G., Brownell, N.: Designing tomorrow: Preparing teachers as change agents for the classroom of the future. Computers in the Schools 8(1-3), 21–24, 45 (1991)
3. Bryant, K., Campbell, J., Kerr, D.: Impact of web based flexible learning on academic performance in information systems. Journal of Information Systems Education 14(1), 41–50 (2003)
4. Clouse, R.W., Alexander, E.: Classroom of the 21st century: Teacher competence confidence and collaboration. Educational Technology System 26(2), 97–112 (1998)
5. Dexter, S.L., Anderson, R.E., Becker, H.J.: Teachers' views of computers as catalysts for changes in their teaching practice. Journal of Research on Computing in Education 31(3), 221–239 (1999)
6. Dias, L.B.: Integrating technology: Some things you should know. Learning & Leading with Technology 27(3), 10–13, 21 (1999)

7. Ertmer, P.A., Addison, P., Lane, M., Ross, E., Woods, D.: Examining teacher, beliefs about the role of technology in the elementary classroom. Journal of Research on Computing in Education 32(1), 54–72 (1999)
8. Ministry of Education. University Distance Learning Implementation Measures. Computer center of Ministry of Education, Taiwan (2011)
9. Ministry of Education. Review the results of teacher selection. Central Region Office of Ministry of Education, Taiwan (2010)
10. Moersch, C.: Levels of technology implementation: A framework for measuring classroom technology use. Learning and Leading with Technology 23(3), 40–42 (1995)
11. Soner, Y.: Effects of an educational computing course on pre-service and in-service teachers: A discussion and analysis of attitudes and use. Journal of Research on Computing in Education 32(4), 479–492 (2000)

Analysing Students' Use of Recorded Lectures through Methodological Triangulation

Pierre Gorissen[1], Jan van Bruggen[2], and Wim Jochems[3]

[1] Fontys University of Applied Sciences, The Netherlands
P.Gorissen@fontys.nl
[2] Fontys University of Applied Sciences, The Netherlands
and
Open University of The Netherlands
J.vanBruggen@fontys.nl
[3] Open University of The Netherlands
Wim.Jochems@ou.nl

Abstract. Recorded lectures provide an integral recording of live lectures, enabling students to review those lecture at their own pace and whenever they want. Most research into the use of recorded lectures by students has been done by using surveys or interviews. Our research combines this data with data logged by the recording system. We will present the two data collections and cover areas where the data can be triangulated to increase the credibility of the results or to question the student responses. The results of the triangulation show its value, in that it identifies discrepancies in the students' responses in particular where it concerns their perceptions of the amount of use of the recorded lectures. It also shows that we lack data for a number of other areas. We will still need surveys and interviews to get a complete picture.

1 Introduction

The lecture method has been around since before the time of the printed book, when monks would read out a book, at a lectern, and the scholars would copy down what was said word for word [8]. Even now it is the most commonly used form of teaching in universities around the world [5]. More and more universities, however, have started to create recordings of these lectures [14], allowing students to review lectures at their own pace and at a time and place of their choosing.

The increase of the number of recorded lectures has been made possible by the availability of more advanced Lecture Capture Systems (LCS). A LCS handles the simultaneous capture of both the audio and video of the lecturer and everything that is being projected during the lecture, usually a PowerPoint presentation. It handles the automatic synchronisation of all the captured media, uploads the recorded lecture to a server and can post a link to the recording in the Virtual Learning

L. Uden et al. (Eds.): Workshop on LTEC 2012, AISC 173, pp. 145–156.
springerlink.com

Environment, notifying students the recording is available. The students can then view the recording in a web browser. Figure 1 shows an example of a recorded lecture with both the video of the presenter and a view of the projected PowerPoint slide side by side.

Fig. 1 Example of a recorded lecture

Much of the existing research into recorded lectures has been focused on improvements of the technology used to record the lectures. Researchers tried to improve the quality of the recordings by addding more advanced interaction options [3, 4], automated capturing [6, 1, 27] and camera control [13], search options [11] and mobile solutions [16]. More recently there also has been more focus on the use of the recorded lectures by students [23, 9, 21], their use in university settings [28, 17], its use for students with a handicap [18] and possible impact of recorded lectures on the attendance of students [26, 22]. Little is known, still, about the way in which students navigate within the recordings or how they find (the parts of) the recordings they want to watch.

The goal of this research is to get a better understanding of how students use recorded lectures and how we can help them to navigate more efficiently to the parts of the recordings they want to view. The main research questions for our study are:

- How do students use recorded lectures?
  - How do students say they use recorded lectures?
  - What actual usage of the recorded lectures can we derive from the data on the system and does that match with what students say?

- What usage patterns can we identify in both the reported and actual usage of recorded lectures by students?
- How can we best facilitate the usage of recorded lectures by students?

Much of the existing research in this area is based on surveys and verbal reports by students of their use of recorded lectures. Triangulation [7] has been successfully used to compare survey data and log file data of Learning Management Systems [15, 19]. In this paper we look at methodological triangulation as a way to increase the credibility and validity of the collected survey data by combining them with data logged by the LCS. We will present the two data collection methods and we will look at the areas where we can triangulate the collected data. What data can be provided by the LCS log data and do we still need surveys to ask students about their use of recorded lectures?

2 Method

For our research we used two methods of data collection: we conducted an online survey combined with semi-structured interviews to collect verbal reports by students. Our second source of data consists of log data that is collected by the LCS.

Participants in the survey were students from various faculties of the Eindhoven University of Technology (TU/e). The TU/e uses the Mediasite LCS to create recorded lectures. A total of six courses that had taken place recently were selected for the survey.

The first part of the survey asked students for their interest in the topic of the course, the perceived importance of the course for their course of study and the grade they wished to achieve for the course. In the second part of the survey, students rated the effectiveness of a number of available activities (e.g., attending face-to-face lectures) and supporting resources (e.g., slides, lecture notes, etc.) in helping them to succeed in the course. It also asked about any previous experience with lecture recordings, and whether they had used the recorded lectures for the course in question. In part three of the survey, those students who had used the lecture recordings were surveyed in more detail about their experiences during that use. Those questions were not displayed to students that indicated they had not used the recorded lectures. The final part of the survey contained questions for all students, seeking out reasons they did not watch one or more of the recorded lectures (if applicable).

The courses all had a set of lectures that were recorded on a regular basis (weekly or more often). Most of the recordings are traditional university-style lectures with the teacher standing in front of the class lecturing. Exceptions to this were lectures where assignments and the test were discussed. All recordings are between 40-45 minutes long. In all of the recordings, video of the lecturer is recorded and displayed.

Five of the courses used PowerPoint or other computer-based applications recorded alongside the video of the lecturer. Two of the courses contained recordings of the lecturer and the blackboard. Table 1 shows the selected courses for the survey, the department responsible for the courses, the number of students per course, the response rate per course for the survey, what was being recorded and the number of recordings per course.

Table 1 Courses selected for the survey and response rates

Course	Department	N	Responses		What is being recorded?		#[1]
			n	(%)	PowerPoint	Blackboard	
C01 Methods and models in behavioural research	Industrial Engineering & Innovation Sciences	307	144	45.6	Yes[2]	Yes[3]	35
C02 Control Systems Technology	Mechanical Engineering	190	72	34.7	Yes[4]	Yes	20
C03 Chemical Biology	Biomedical Engineering	136	68	49.3	Yes	Yes	27
C04 Facades and Roofs	Architecture, Building and Planning	115	40	33.9	Yes	No	15
C05 Vector calculus	Applied Mathematics	94	47	48.9	No	Yes	14
C06 Calculus	Applied Mathematics	77	43	55.8	No	Yes	35

[1] Number of recordings for this course
[2] Both PowerPoint and demos of applications
[3] For additional notes, during 5 recordings
[4] During 8 of the 20 recordings

Student selection for the survey was based on recent participation in one of the seven selected courses. All students in the courses had a choice between either attending the lecture, viewing it online, or doing both. We approached the students using a personalized e-mail that contained the link to the web-based survey. In the e-mail and the survey itself, the students were asked to complete the survey based on their experiences and use for the one specific course for which they were selected. The survey was open online for two weeks. An e-mail reminder was sent after one week and again on the final day of the survey to those students who had not completed the survey.

The online survey contained seventeen questions using both multiple choice and Likert scale questions. Some of the questions have been used in other surveys on the use of recorded lectures [10, 12, 21, 23, 24, 25, 28]. Students were able to complete the survey in about 10-15 minutes. As part of the survey, we invited students for follow-up questions. A total of 120 students accepted the invitation initially.

Of those students, 14 were interviewed using a semi-structured interview lasting 30 minutes. During the interviews, students were asked to elaborate on their use of the recordings during the course. The interviews were recorded and transcribed.

The second data set contains data collected by the LCS. All recordings are available online; students can view them in their browser, both at the university and from home. Students need to login using their university account to view the recorded lectures. No downloadable versions of the recordings are provided. Whenever someone views a recorded lecture, a log entry is created by the system detailing the time and date of the view, the recorded lecture that was viewed, the user that viewed the recorded lecture and the parts of the recorded lecture that were sent to the user.

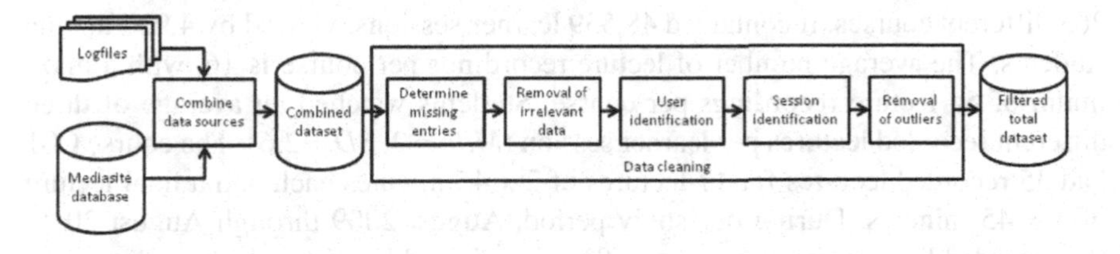

Fig. 2 Data pre-processing steps

We performed a process called 'data pre-processing' [20] to prepare the data set for analysis. Figure 2 shows the steps taken during this process. The data from the Mediasite LCS were available in the database and text-based log files. We combined the two data sources and then did a number of data cleaning steps. We removed the data for all users other than students. This included other staff, the professors/lecturers, and the researchers conducting this analysis. Also, we grouped the interactions of students into learner sessions. A learner session is an uninterrupted period of time during which a learner accesses one or more recorded lectures [2]. This does not mean that a learner session consists of a constant viewing of the recorded lectures. We assume that students also review notes, do assignments, read in their textbooks, take a short break, etc. during a learner session. Finally, the data cleaning removed all outliers from the data set. We were only interested in learner sessions where the students actually make use of the recorded lectures. Learner sessions shorter than three minutes or learner sessions where a total of less than 2 minutes of video has been received by the student are not considered to be actual use of the recorded lectures as part of study activities and were removed from the dataset.

The analysis of the actual use of the recorded lectures focused on one of the courses selected for the original survey and interviews. This course, C01, is a course at the Industrial Engineering & Innovation Sciences department of the TU/e. Students that participate in the course come from a number of different departments within the university. Most of the students (66%) are from the Industrial Engineering department; another substantial group of students (23%) is from the Innovation Sciences department. The course consists of an introduction in empirical research.

Students learn how to translate real-life questions into research questions, and they learn how to create and evaluate a research design. In the second part of the course, they get hands-on training using SPSS.

3 Results

The total response rate for the survey conducted at the TU/e was 414 (45.1%, N = 919); the response rate for course C01 was 143 (46.6%, N = 307). For the remainder of this results section, we will report on the students for the C01 course only.

The filtered total dataset contained data on 4,192 lecture recordings, for a total of 263 different courses. It contained 48,539 learner sessions, viewed by 4,927 unique students. The average number of lecture recordings per course is 16, with a maximum of 54 lecture recordings per course. Students watched an average of three different recorded lectures per learner session (*Mdn* = 2, *SD* = 2.6). The course C01 had 35 recorded lectures for 17 lectures of 2 x 45 minutes each and a final lecture of 1 x 45 minutes. During our study period, August 2009 through August 2010, the recorded lectures for the course C01 were viewed by 291 unique students in a total of 2,650 learner sessions. The surveys and the follow up interviews provide contextual information about the students using the recorded lectures created for the course C01, not available in the log data collected by the LCS. Most students felt that the topic of the course was important (66.7%) and agreed that the course was an important part of their study (68.5%). On average, students aimed for a 7 (on a 10 point scale) as the grade for this course. Students rate lecture recordings almost as high as attending face-to-face lectures when asked about their effectiveness in helping them to succeed in the course. Both score considerably higher than the online virtual learning environment used for the course or the help of other students.

The survey asked the students about technical difficulties while watching the recorded lectures. Most students (47.2%) reported that there were no technical difficulties, 20.8% mentioned slides and video not always playing synchronous, and 14.6% reported bad video quality. The log data from the LCS does contain some information about the bandwidth used during the playback of the recorded lecture and possible lost packets of data sent to the student, but that cannot be translated into real technical difficulties like shown in the table.

We also asked student to indicate the importance of a number of features available in the player for the recorded lectures. Table 2 shows their responses to that question. Table 2 shows that replaying the recorded lecture at higher or lower speed, navigating through the recorded lecture using the slide list and using the play head are features found important by a majority of the students. Although it would be possible to track the use of the above mentioned features, there is currently no data about actual use in the LCS log data to substantiate or correct these reports. The player used to display the recorded lectures does not send any information related to the method of navigating or the speed at which the video is displayed back to the server.

Table 2 Features indicated as somewhat or very important while viewing recorded lectures

	n	(%)
Playing at higher or lower speed	110	85.2
Navigating using the slide list	98	76.0
Scanning through video using play head	80	62.0
Muting sound / controlling sound level	78	60.4
Skipping back	75	58.1
Viewing the lecture recording offline	48	37.2
Downloading additional resources via presentation links	39	30.3
Saving links to specific locations in the lecture recording	34	26.4
Mailing questions to lecturer from within viewer	9	7.0
Sharing lecture recording via mail with other students	9	7.0

There are also questions in the survey that can be linked to LCS log data, either directly or indirectly. In the survey, we asked students how important different purposes of recorded lectures were to them. Table 3 shows that making up for a missed lecture and preparing for the exam score highest for course C01 while also the number three in the list, improving test scores, can be seen as an indication that preparing for the exam is an important use of the recorded lectures for students.

We compared the responses of the students for the two most important purposes (making up for a missed lecture and preparing for the exam) with the data available in the LCS logs.

We cannot directly link the viewing of a recorded lecture to an actually missed lecture. Lecture attendance was not mandatory and no attendance register was created during the C01 course for 2009-2010. Instead, we assume that if students used recorded lectures as a substitute for lecture attendance, they would watch the full length of a recorded lecture. We assume that they have watched the full length of a recorded lecture if they received at least 80% of the video for the recorded lecture.

On average, each recorded lecture for the C01 course is watched in full for 11 times during the one-year period covered by the dataset. The maximum number of recorded lectures watched in full, by a single student for the C01 course, is 20 recorded lectures out of a total of 34 successfully recorded lectures. There were 13 students who watched 10 or more recorded lectures in full. Most of the students never watched the full length of a recorded lecture for this class. Of all students, only 27% watched the full length of one or more of the recorded lectures. So, although 96.2% of the students consider "making up for a missed lecture" an important purpose of the recorded lecture, the LCS log data does not support this response because most of them never ever watched the full length of a recording. We analysed the number of learner sessions for the course C01 based on the LCS log data.

Table 3 Somewhat or very important purposes of using recorded lectures

	n	(%)
Making up for a missed lecture	124	96.2
Preparing for the exam	120	93.0
Improving test scores	112	86.8
Improving retention of lecture materials	102	79.1
Clarifying the material	99	76.8
Replacing live attendance	96	74.4
Assisting with an assignment	88	68.2
Reviewing material after a lecture	70	54.3
Managing distractions during lectures	64	49.6
Reinforcing the experience at the live lecture	46	35.6
Reviewing material before a lecture	43	33.4
Checking own notes	33	25.6
Overcoming language barriers	16	12.4

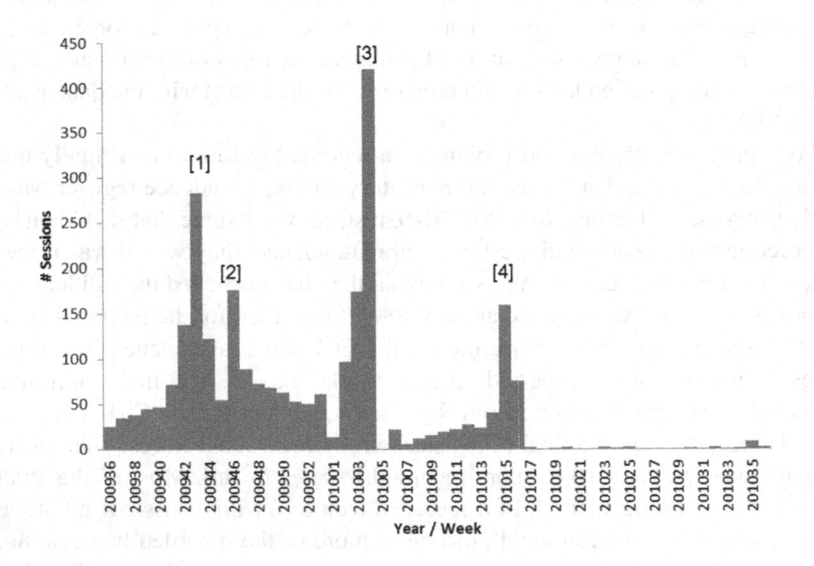

Fig. 3 Number of learner sessions per week

Table 4 Number of times respondents used recorded lectures for the C01 course

	Reported		Actual	
	n	(%)	n	(%)
Never	13	9.1	6	4.2
<5 times	22	15.4	35	24.5
5-10 times	51	35.7	43	30.1
>10 times	57	39.9	59	41.3

Table 5 Average percentage of a recording viewed

	Reported		Actual	
	n	(%)	n	(%)
0% - 10%	2	1.5	27	9.3
10% - 25%	4	3.1	203	69.8
25% - 50%	7	5.4	40	13.7
50% - 75%	26	20.0	13	4.5
75% - 100%	91	70.0	8	2.7

Figure 3 shows the number of learner sessions for course C01 per week. It shows that there are four weeks in which there is an above-average use of the recorded lectures, indicated as [1] – [4] in Figure 3. These are the weeks that the assignment for the course is due, a written test is planned, the laptop test is scheduled and the week before the retest takes place. In this case, the log data does support the responses by the students. The recorded lectures are used a lot during the preparation for the exams.

There are two further questions further questions in the survey for which we can use triangulation with the LCS log data. Table 4 shows the number of times respondents indicated they had used recorded lectures for the course, compared to the actual number of learner sessions based on the LCS log data for the students that completed the survey for course C01. The table shows that the reported number of learner sessions for this course is approximately equal to the actual measured numbers of learner sessions. This cannot be said about the reported and actual percentage of a recorded lecture that students view on average. Table 5 shows that the majority of students say that on average they watch three quarters or more of a recorded lecture. When measuring the average amount of video of a recorded lecture that is actually sent to a student, the vast majority only receives between 10% - 25% of a recorded lecture.

4 Conclusions

The analysis shows that the survey still is an important method to collect information from students about their use of recorded lectures. The data logged by the LCS does not provide all the information that we want and need to get the complete picture. However, methodological triangulation is a valuable step to confirm or to question at least some of the students' responses. It is not sufficient to rely on just the self-reported data by students. The triangulation confirmed the reported number of learner sessions for this course but also showed that the percentage of a recorded lecture that students report to watch is much higher than the actual percentage. The analysis also shows that sometimes there is a difference between what students report to be an important purpose for recorded lectures and their actual use of the recorded lectures.

In some cases where the data logged by the LCS currently is insufficient to perform triangulation, improvements can be made. For example, the methods that a student uses to navigate through the player interface is not yet logged, but could provide valuable information about whether the interface allows them to quickly find the parts of a recorded lecture that they want to view.

To be able to rely on this combination of datasets, unique identification of users is very important. We've seen examples where a single recording had been viewed on the same university computer by three different students on a single day. Just counting IP-addresses would provide incorrect information. Downloadable recorded lectures can also not be counted or tracked. Universities should be aware of those effects on the completeness of reports that they can create.

Acknowledgements. This research has been funded by Fontys University of Applied Sciences. The authors would like to thank the Student Office of the TU/e for their assistance in this research.

References

1. Abowd, G., Atkeson, C., Brotherton, J., Enqvist, T., Gulley, P., LeMon, J.: Investigating the capture, integration and access problem of ubiquitous computing in an educational setting. Paper Presented at the SIGCHI Conference on Human Factors in Computing Systems, Los Angeles, California, United States (1998)
2. Advanced Distributed Learning, SCORM Run-Time Environment Version 1.3 (2004), `http://www.adlnet.org/` (accessed September 16, 2010)
3. Arons, B.: Speech Skimmer: a system for interactively skimming recorded speech. ACM Transactions on Computer-Human Interaction (TOCHI) 4(1), 3–38 (1997)
4. Baecker, R.M., Moore, G., Zijdemans, A.: Reinventing the lecture: webcasting made interactive. In: Stephanidis, C. (ed.) Proceedings of HCI International 2003, pp. 896–900. Lawrence Erlbaum Associates, New Jersey (2003)
5. Behr, A.L.: Exploring the lecture method: An empirical study. Studies in Higher Education 13(2), 189–200 (1988)

6. Brotherton, J., Abowd, G.: Lessons learned from eClass: Assessing automated capture and access in the classroom. ACM Transactions on Computer-Human Interaction (TOCHI) 11(2), 121–155 (2004)
7. Denzin, N.: Sociological Methods: A Sourcebook, 5th edn. Transaction Publishers, Piscataway (2006)
8. Exley, K., Dennick, R.: Giving a lecture: from presenting to teaching. Routledge/Falmer, London (2004)
9. Filius, R.: De huiskamer als cursuslokaal, flexibel leren met weblectures [The living room as a lecture hall, flexible learning using weblectures]. Develop. 4, 30–41 (2008)
10. Hall, D.: My Media Student Evaluation 2009. Centre for Learning and Professional Development. University of Adelaide, Adelaide (2009)
11. Hürst, W.: Indexing, searching, and skimming of multimedia documents containing recorded lectures and live presentations. In: Rowe, L., Vin, H., Plagemann, T., Shenoy, P., Smith, J. (eds.) Eleventh ACM International Conference on Multimedia, pp. 450–451. ACM, Berkeley (2003)
12. Kishi, C., Traphagan, T.: Lecture Webcasting at the University of Texas at Austin. In: Proceedings of the 23rd Annual Conference on Distance Teaching & Learning, pp. 1–5. University of Wisconsin, Madison (2007)
13. Lampi, F., Kopf, S., Benz, M., Effelsberg, W.: An automatic cameraman in a lecture recording system. Paper Presented at the International Workshop on Educational Multimedia and Multimedia Education, Augsburg, Bavaria, Germany (2007)
14. Leoni, K., Lichti, S.: Lecture Capture in Higher Education. Northwestern University, Evaston (2009)
15. Lonn, S., Teasley, S.: Saving time or innovating practice: Investigating perceptions and uses of Learning Management Systems. Computers & Education 53(3), 686–694 (2009)
16. Read, B.: Lectures on the Go. In: The Chronicle of Higher Education, October 28 (2005)
17. Russell, K., Fass, H., Bloothooft, G.: Rapportage project Weblectures [report on project Weblectures]. Utrecht University, Utrecht (2008)
18. Russell, K., Filius, R., te Pas, S.: Verslag Grassroots project Opnemen en uitzenden van hoorcolleges voor studenten met een handicap [Report on Grassroots project Recording and Broadcasting lectures for students with a disability]. Utrecht University, Utrecht (2007)
19. Sheard, J.: An Investigation of Student Behaviour in Web-based Learning Environments. PhD diss. Monash University, Victoria, Australia (2007)
20. Sheard, J.: Basics of Statistical Analysis of Interactions. Data from Web-Based Learning Environments. In: Romero, C., Ventura, S., Pechenizkiy, M., Baker, R.S.J.D. (eds.) Handbook of Educational Data Mining, pp. 27–40. Chapmann & Hall/CRC Press, Boca Raton (2011)
21. Traphagan, T.: Class Lecture Webcasting, Fall 2004, Spring 2005, and Fall 2005: Summary of Three Case Studies. University of Texas, Austin (2006)
22. Traphagan, T., Kucsera, J., Kishi, K.: Impact of class lecture webcasting on attendance and learning. Educational Technology Research and Development 58(1), 19–37 (2010)
23. Veeramani, R., Bradly, S.: Insights regarding undergraduate preference for lecture capture. University of Wisconsin-Madison E-Business Institute, Madison (2008)
24. Wieling, M.: De effecten van het aanbieden van videocolleges als aanvulling op de reguliere hoorcolleges binnen de Faculteit Rechten [The effects of recorded lectures as a supplement to regular lectures at the Law department]. Groningen University, Groningen (2008)

25. Williams, J., Fardon, M.: Perpetual Connectivity: Lecture Recordings and Portable Media Players. In: Atkinson, R., McBeath, C., Soong, S., Cheers, C. (eds.) Proceedings Ascilite Singapore 2007, pp. 1084–1092. Centre for Educational Development, Nanyang Technological University, Singapore (2007)
26. Williams, J., Fardon, M.: Recording lectures and the impact on student attendance. Paper Presented at the ALT-C 2007 Conference (2007b)
27. Zhang, C., Crawford, J., Rui, Y., He, L.: An automated end-to-end lecture capturing and broadcasting system. Paper Presented at the 13th Annual ACM International Conference on Multimedia, Singapore (2005)
28. Zupancic, B.: Vorlesungsaufzeichnungen und digitale Annotationen: Einsatz und Nutzen in der Lehre [Lecture recording and digital annotations: Use and Benefits in Teaching]. PhD diss., Universität Freiburg, Fakultät für Angewandte Wissenschaften, Freiburg (2006)

A Study on the Effect of Constructivism-based Teaching on Industrial High School's Electronics Practice

Chin-Wen Liao, Hsuan-Lien Chen, Chen-Jung Lai, and Chih-Hao Chen

Department of Industrial Education and Technology,
National Chunghua University of Education. No.2, Shi-da road,
Chunghua City, 500, Taiwan, R.O.C.
tcwliao@cc.ncue.edu.tw, hsuan.lien@msa.hinet.net,
tcivs74@yahoo.com.tw, chenuj22@hotmail.com

Abstract. The objective of this study was to examine the effect of implementing constructivism-based teaching on electronics practice in industrial high schools in terms of students' practice achievement, learning motivation and post-learning attitude. Suggestions were formulated based on the research findings for teachers to take into consideration.

This study adopted the experimental research method using nonequivalent pretest-posttest control group design. A total of 120 students participated in this study. They were from three second-year electronics-major classes (Class A, B and C) in National Taichung Industrial High School. The study subjects were divided into two groups: control vs. experimental. The experimental group was taught by constructivism-based teaching approach for fourteen weeks (forty-two classes), while the control group was taught by the conventional teaching approach.

The research findings according to the above-mentioned research objectives are:

1) Electronics practice learning achievement: Results from one-way ANOVA suggested that the difference between the experimental and the control group in their electronics practice exam grades was not statistically significant.
2) Electronics practice learning motivation: Results from one-way ANOVA suggested that students from the experimental group scored higher than those from the control group did in electronics practice learning motivation.
3) Electronics practice post-learning attitude: The mean and the standard deviation of the scores of students from the experimental group ranged between *neither satisfied nor dissatisfied* and *satisfied* in the cognitive subscales, the technological subscale and the affective subscale.

Keywords: Constructivism-based teaching, Electronics practice learning achievement, Electronics practice learning motivation.

L. Uden et al. (Eds.): Workshop on LTEC 2012, AISC 173, pp. 157–166.
springerlink.com

1 Introduction

1.1 Research Background

To attain education reform, schools should help students develop a variety of skills, including problem solving, communication, critical thinking, collaborative learning, interpersonal development and decision-making (Fan, 1999). In order to achieve the above-mentioned education goals, the conventional education mindset should be revised. That is, instead of considering students as solely the knowledge receivers while their teachers as the center of the teaching-learning process, pedagogical approaches focused on analytical thinking, flexibility and problem-solving should be adopted to enhance students' continuing education capacity and to elicit their learning motivation. This new approach can also create a livelier learning environment that replaces competition by collaboration. Ultimately, an interest in lifelong learning and development can be promoted.

Therefore, this study adopted the experimental research approach, analyzed the literature to design a Constructivism-based teaching approach suitable for industrial high schools' electronics practice, and examined the effect of this approach on students' learning achievement, learning motivation and post-learning attitude.

1.2 Objectives

The objective of this study is to use the electronics practice teaching procedure to understand students' entering behavior. Constructivism-based teaching approach was utilized to review the implementation of the teaching and to revise the pluralistic teaching strategies. Impacts from Constructivism-based teaching on industrial high school students taking the electronics practice course were evaluated by students' pre- and post-experiment tests on electronics practice, their learning motivation, and their post-learning attitude.

1.3 Questions to Be Answered

The following questions were formulated based on the above-mentioned research questions:

1) Is there a significantly different level of learning achievement in industrial high school electronics practice between students taught by constructivism-based approach and students not taught by Constructivism-based approach?

2) Is there a significantly different level of motivation in electronics practice between students taught by Constructivism-based approach and students not taught by Constructivism-based approach?

3) How do the industrial high school students taught by Constructivism-based approach score in the post-learning attitude test after completing the electronics practice?

2 Problem Description and Background

2.1 *Basic Theory of Constructivism*

Constructivism is an epistemological perspective considering that the acquisition of new knowledge is based on students' voluntary knowledge building and reconstructing of knowledge about the external world. Moreover, the type of knowledge being generated is determined by reflection, cross-examination, and actions (Yu, 1997). When interacting with the external world, learners would judge and learn about the surrounding environment based on the knowledge they possessed. When the available knowledge is inefficient for coping the external world, learners would adjust their presently owned knowledge system in order to deal with the actual condition. Therefore, according to Constructivism, the formation of knowledge is an active construction instead of a passive process of acceptance.

2.1.1 Constructivism Teaching Strategies

1) Active knowledge acquisition includes: a) Developing voluntary-type of assignments; b) Adopting cognitively guided instruction (CGI); c) Implementing computer network-based teaching; d) Teaching students about learning strategies; e) Stimulating students' desire for learning; f) Making questions the focus in class; g) Using concept map for clarifying problems; h) Writing journals for stimulating multidimensional wisdom.

2) Interactive learning includes: a) Arranging conversation between students and teachers as well as among the students themselves; b) Letting students observe lecture delivery; c) Carrying out cognitive interactive teaching; d) Establishing cognitive apprenticeship; e) Adopting collaborative learning; f) Sharing brainstorming results; g) Encouraging instant interactive and peer sharing.

3) Multidimensional experience includes: a) Adopting a context-simulating teaching approach; b) Enhancing pedagogical content knowledge; c) Arranging the do-it-and-learn opportunities; d) Applying multidimensional teaching strategies; e) Adopting pluralistic and realistic evaluation.

2.1.2 Constructivism's Pedagogical Model

American education psychologist G. Glaser (1962) proposed a general model of instruction (GMI), which includes pedagogical objectives, initiative behavior, pedagogical portfolio, and pedagogical evaluation. After reviewing works of other experts, the researchers adopted the Constructivism-based teaching approach, and the procedures and process are presented below:

Table 1. Procedures for Constructivism-based teaching approach

Step	Step 1	Step 2	Step 3	Step 4	Step 5	Step 6
Teaching Activities	Pilot learning	Demonstration and explanation	Active in-spection	Group learning	Discussion and clarification	Multidimensional evaluation

3 Experimental Design and Implementation

3.1 Research Structure

3.1.1 Independent Variables

Pedagogical approach: The experimental group was taught by Constructivism-based approach three hour per week for fourteen weeks (forty-two hours). The teaching activity includes six steps, which were pilot learning, explanation and demonstration, voluntary investigation, group learning, discussion and explanation, and pluralistic evaluation. The control group, on the other hand, was taught by the conventional approach.

3.1.2 Dependent Variables

1) Electronics practice learning achievement: This aspect was assessed by students' grades from the midterm and final exams of the course *Electronics Practice*. The exam was developed by the researchers.
2) Electronics practice learning motivation: This aspect was assessed by *Motivational Belief Scale* developed by Chang (1994). There are three subscales, which are *self-efficacy, internal value*, and *achievement motivation*.
3) Students' post-learning attitude: This aspect was assessed by *Students' Post-Learning Attitude Scale for Constructivism-Based Teaching*, which was developed by the researchers.

3.1.3 Control Variables

1) Students from three electronics-major classes of National Taichung Industrial High School had similar school entrance grades when admitted into the school. Their class placement was determined by their last name. In other words, the levels of students from these three classes were roughly similar and were unlikely to confound the experimental results.
2) Grades from the pre-experiment *Electronics Practice* exam were treated as a covariate to prevent students' prior capability influencing the experimental results.

3) The pre-experiment learning motivation score was treated as a covariate to prevent students' prior learning motivation affecting the experimental results.

3.1.4 Research Tools

Electronics Practice tests given prior to the beginning of the course and at the midterm as well as the final exams were developed according to the following procedures. First, a learning achievement scale was set up. Second, for the validity of the test, four professors from graduate schools and three teachers teaching *Electronic Practice* at the Taichung Industrial High School were asked to revise the test questions. Third, a pretest was hold. For general tests, the difficulty level index should be between 0.20 and 0.80, and the most optimal level should be close to 0.50. The accepted lowest differentiation index, on the other hand, should be 0.25 or above. Fourth, in terms of the reliability, the index has to be greater than 0.65 for group comparison. The test designed for the *Electronics Practice* had a reliability of 0.76, and therefore, it met the standard.

The learning motivation scale used for *Electronics Practice* was modified from *Motivational Belief Scale* developed by Chang (1994).

The post-learning attitude scale was developed by the following two steps. First, questions from the pretest scale were distributed. Secondly, the pretest questions were analyzed for the following areas: For item analysis, *t*-test was used to eliminate questions with a critical ratio that did not reach the significant level. For correlation analysis, questions with a correlation less than 0.30 or was insignificant were eliminated. Reliability analysis was conducted before eliminating the questions. Factor analysis was adopted for creating the validity of the scales. Principal component extraction was used for the remaining questions, and then direct oblimin was applied to divide the questions into three factors, with 0.30 as the factor loading critical value. Reliability analysis was again conducted after eliminating the questions.

3.2 Designing the Constructivism-based Teaching Activity

The teaching activity of this study can be divided into six steps.

The first step was pilot learning, and it included the followings. (1) What students already know was assessed based on student-teacher interaction to identify students' entering behavior. (2) Key points from the previous lesson were reviewed and connected to the key points of the present lesson. (3) Students were asked to write the *Pre-lesson Self-examination Questionnaire*. (4) Key points of this lesson were presented to motivate students to learn the content of this lesson.

The second step was to give demonstration and explanation, and it included the followings. (1) Media that were stimulating, novel and variable were adopted to facilitate learning and maintain students' attention. (2) Rhe content of the course was organized and extracted to facilitate learning and to construct the focus of the course.

The third step was voluntary exploration, and the followings are the key points. (1) The practice items were closely link to real life situations. (2) Students were guided to think, to construct the knowledge independently, and to clarify the essence of the question (Yeh, 1998).

The fourth step was about group learning. (1) Collaborative learning and group discussion was used to ask the more advanced students to help those lag behind. (2) Each group had four students, and students were asked to practice within their group. (3) The group reward and individual accountability approach were adopted. Students were asked to help others in their group.

The fifth step was discussion and explanation. (1) To share information with other groups, each group assigned one person to talk about the implementation process. (2) The student presentation enabled students to recall steps in the practice. This enlarged the depth and the width of learning. (3) Students were asked to observe other people's problem solving. This can elevate their strategic thinking and promote problem-solving reflection.

The sixth step was pluralistic evaluation, and the key points are as follows. (1) Students were asked not to repeat the practice robot-like. Moreover, the best group was asked to give demonstration and exercise. (2) The evaluation was made to be realistic, and pluralistic problem-solving strategies were the focus. (3) Besides problem-solving outcomes, the calculation and deduction process were emphasized as well. (4) Note taking was encouraged. Students were asked to write down ideas and comments about learning as well as the learning content. This can help teachers understand students' response to the lesson.

4 Results and Conclusions

4.1 Discussion on the Effect of Constructivism-based Teaching on Students' Electronics Practice Achievement

Grades from Week 1, the midterm and final *Electronics Practice* exams were analyzed by the homogeneity and regression test, but no significant difference was found ($F=0.069$, $P>0.05$). Covariance analysis was further conducted. After eliminating the effect on the two groups' Week 1 *Electronics Practice* test grades, the obtained result suggested no significant difference between the two groups in the midterm and final exam grades ($F = 0.24$, $P > 0.05$). In other words, the Constructivism-based teaching approach did not improve students' electronics practice achievement.

Table 2 The covariance analysis table for the two groups' Electronics Practice test grades

Source of variance	Sum of squared deviations from the mean	Degree of freedom	Mean square	F-value
Between-group	23.70	1	23.70	.24
Error	3484.66	35	99.56	

Possible reasons are as follows:

1. Students were unfamiliar with constructivism-based teaching strategies.
2. The timing of the achievement tests was inappropriate. Usually, when Constructivism-based researchers give this type of tests, they would give the tests right after the experimental group has completed the lesson. They do so because it can reduce the moderating variables to the minimum. In this study, however, the tests were not given immediately after the completion of the lesson. As a result, students' performance may be affected by moderating variables.

4.2 Discussion on the Effect of Constructivism-based Teaching on Students' Learning Motivation

The regression and homogeneity analysis on the two groups' scores from Week 1's *Electronics Practice Learning Motivation Scale* (pretest) and *Electronics Practice Learning Motivation Scale* (posttest) suggested no significant difference ($F = 1.11$, $P > 0.05$). Covariant analysis was performed. After eliminating influences from the pretest score of *Electronics Practice Learning Motivation Scale*, a significant difference was found between the two groups in the posttest score of *Electronics Practice Learning Motivation Scale* ($F = 12.04$, $P < 0.05$).

Possible reasons are:

1) Students' strong impression: Constructivism-based teaching approach is different from the conventional teaching approach. It includes pilot learning, explanation and demonstration, voluntary exploration, group learning, discussion and explanation and pluralistic evaluation. Because the students had never been exposed to this type of teaching approach, they were more likely to be strongly impressed by this type of teaching.

Table 3 The covarince analysis table for the two groups' scores from *Electronic Practice Learning Motivation Scale*

Source of variance	Sum of squared deviations from the mean	Degree of freedom	Mean square	F-value
Between-group	1353.74	1	1353.74	12.04*
Error	3936.47	35	112.47	

2) The researchers adopted the interactive teaching approach. This approach stresses that learning is an internalization process through social interaction. When using this approach, the instructor would first demonstrate the practice, while the learners observe. Gradually, the learners will participate in more and more activities. In this approach, group reward and individual accountability are applied, and each one should help others.
3) Role model can facilitate the learning process: In the teaching activity, each group had the more advanced student sharing his or her learning process with other. These students acted as the role model for those lag behind and helped others improve their learning capability.

4.3 Students' Perspective Based on Their Post-Electronics Practice Learning Attitude

4.3.1 Cognitive Learning

The experimental groups' average score of satisfaction after Constructivism-based teaching was between 2.57 and 2.96, or between *neither satisfied nor dissatisfied* and *satisfied*. Students from the experimental group considered that Constructivism-based teaching can elicit students' learning motivation and teach students many novel concepts, and thus it is helpful. They also thought that this approach could be used for other professional courses.

4.3.2 Skill Learning

The experimental groups' average satisfaction score after Constructivism-based teaching was between 2.62 and 2.85, or between *neither satisfied nor dissatisfied* and *satisfied*. Students from the experimental group considered that Constructivism-based teaching could reduce their work time for the practice and facilitate the electronic circuit troubleshooting. Therefore, they think it is a good learning approach.

4.3.3 Affective Domain

The experimental groups' average satisfaction score after Constructivism-based teaching was between 2.52 and 3.01, or between *neither satisfied nor dissatisfied* and *satisfied*. Students from the experimental group considered that Constructivism-based teaching could make them more confident about electronics practice, facilitate the planning of practice work, increase their analytical and judgmental capacity, and elevate their learning efficacy.

5 Conclusion

The research finding suggested:

1) There was no significant difference between the experimental and the control groups in *Electronics Practice*'s midterm and final exam grades.
2) The experimental group scored higher in all the posttests of *Electronics Practice Learning Motivation Scale* than the control group did. In other words, there was a significant improvement in learning motivation in the experimental group after Constructivism-based teaching.
3) According to the cognitive subscale of the attitude scale from post-learning, the experimental group tended to consider that Constructivism-based teaching could elicit students' learning motivation and provide them with many new concepts. These are helpful for *Electronics Practice* and may be applicable for other professional courses as well.
4) According to the skill subscale of the attitude scale from post-learning, the experimental group tended to consider that Constructivism-based teaching can reduce the practice work time and facilitate electronic circuit troubleshooting. Therefore, students considered that a good learning approach.
5) According to the affective subscale of the attitude scale from post-learning, the experimental group tended to consider that Constructivism-based teaching can make them more confident about electronics practice, facilitate practice work planning, improve their analytical and judgmental capacity, and elevate learning efficacy.

5.1 Suggestions

5.1.1 Pedagogical Suggestions

Constructivism-based teaching should be promoted to be used in the course *Electronics Practice*. This study had discovered that Constructivism-based teaching could increase industrial high school students' learning motivation. The center of

Constructivism-based teaching is students; students are encouraged to discover problems and solve the problems during the learning process. Constructivism-based teaching also utilizes interactive teaching to help those who lag behind to observe others' learning approach, which is beneficial for their learning.

5.1.2 Suggestions for Future Research

To have a more in-depth investigation of this topic, qualitative studies are a good option. In this study, quantitative analysis was adopted for evaluating the Constructivism-based teaching efficacy. Nevertheless, this approach cannot provide detailed observational information regarding students' learning condition. Moreover, this approach cannot give insights into difficulties encountered by each student in learning. Therefore, the researchers of this study suggest future investigators to adopt a qualitative approach to examine further the research findings in order to understand students' learning.

References

1. Bruner, J.S.: The Act of Discovery. Harvard Educational Review 31, 21–32 (1961)
2. Vygotsky, L.S.: Mind in Society. Harvard University Press, Cambridge (1978)
3. Chang, C.Y.: The Verification and Application of Junior High School Students' Math Learning Process Integrating Model: A Study on the Analysis of Students' Math Concept Build-up and Pedagogical Strategies for Essay-type Math Question. Master Thesis, Graduate School of the Department of Psychology and Counseling, National Taiwan Normal University (1994)
4. Yu, M.N.: Meaningful Learning: A Study on Concept Mapping. Shinning Culture Publishing Co., Taipei (1997)
5. Yeh, C.H.: A Research Experiment on Constructivism-based Teaching for Grade Seven Math Class. Master Thesis, Graduate School of Education, National Chengchi University (1998)
6. Fan, C.C.: A Study on Taiwan's Accounting Education Reform. Master Thesis, Graduate School of the Department of Business Education, National Changhua University of Education (2000)
7. Chen, H.L.: Teaching Efficacy of Metacognition Strategy in Industrial Senior High School Electronic Labs. Graduate School of the Department of Business Education, National Changhua University of Education (2003)
8. Meyer, D.L.: The poverty of constructivism. Educational Philosophy and Theory 41(3) (2008)
9. Buckley, M.: The structure of justification in political constructivism. Metaphiloso. Phy. 41(5) (2010)
10. Joldersma, C.W.: Ernst von glasersfelds radical constructivism and truth as disclosure. Educational Theory (2011)

Ontology in Adaptive Learning Environment

Mona Laroussi

INSAT, Carthage University, Tunis, Tunisia
Mona.laroussi@insat.rnu.tn

Abstract. Adaptive e-learning systems can be defined as systems offering personalized solutions to suit individual learners' needs. Thus personalized solutions are built with respect to different kinds of adaptation which are content and navigation. The goal of such adaptive environments is not to make learning material accessible but to improve learning by adapting navigation, interaction and content. For this aim, it was necessary to investigate the learner profile and activities modelling. Actually, with new kinds of adaptation induced by mobility needs of learner, the learner profile modelling have to be adapted. We propose a more exhaustive learner model which we call "learner context". Thus, in adaptive learning systems, an adequate modelling approach must be proposed and integrated with respect to learning reusability and interoperability. In this respect, semantic Web technologies especially ontology seem to be an efficient modeling approach and can be used to take adaptation decisions. we intend to explore how ontologies can harmonize different aspects of an adaptive e-learning system. Such ontologies could be used to assist designers and authors in the modelling of contextualized learning or even to directly generate such experiences themselves.

Keywords: ontology, Adaptive learning, user model, user context, learning activity.

1 Introduction

E-Learning is an electronic activity that differs from learner to learner. Most of the Learning environments do not take into account individual aspects of learners, ignoring the different needs that are specific to profiles styles and background.

Personalization is the next step in the evolution of eLearning environments. Learners can have several cognitive styles [Laroussi 01], which makes the efficiency and efficacy of an eLearning environment different with distinct learners.

We present in this paper an approach to eLearning adaptation based on ontology. We developed a learner model that is modelled with ontology, enabling the personalization system to guide the student's learning process. In this paper we present adaptive e-learning environment, ontology in e-learning and our use of ontology. This use focuses on learner context, learning resources and activities.

Indeed, ontology is used to improve modelling mechanisms of various types of knowledge relevant for adaptive e-learning. In this paper, we intend to explore how ontology can harmonize different aspects of an adaptive e-learning system.

L. Uden et al. (Eds.): Workshop on LTEC 2012, AISC 173, pp. 167–177.
springerlink.com

To build adaptive e-learning systems, we identified three types of ontology:

- Learning Activity ontology: describes formally components and structure of learning activities;
- Context ontology: provides means to express formally the knowledge about the learner's and activity's context in which activity is performed;
- Resources-ontology: formal description allowing the Sharing of Learning Resource. Resources had to created, searched and retrieved, based on standardized meta-data.

2 Adaptive Learning Environment

2.1 Definitions

An e-learning environment is considered adaptive if it is capable of: monitoring the activities of its users; interpreting these on the basis of domain-specific models; inferring user requirements and preferences out of the interpreted activities, appropriately representing these in associated models; and, finally, acting upon the available knowledge on its users and the subject matter at hand, to dynamically facilitate the learning process. [Paramythis 04]

Accordingly, adaptive e-learning systems carry out adaptation in accordance with a user model [Laroussi 09] which constitutes the main dimension of adaptation in adaptive e-learning systems.

The User Model is a representation of relevant features of the user such as knowledge state, goals, preferences, interests, background, expertise, competence, learning styles, cognitive traits and other features that enable the system to distinguish different users. What to adapt in e-learning environment?

2.2 Dimensions of Adaptativity

In [Laroussi 01], we define for adaptive educational hypermedia 2 dimensions of adaptativity: navigation and content. Those dimensions are extended in e-learning environments. According to [Paramythis, 04], during an e-learning session we can adapt: interaction, course delivery, content discovery and assembly, and, finally, collaboration support.

Adaptive Interaction refers to adaptations that take place at the system's interface and are intended to facilitate or support the user's interaction with the system

Adaptive Course Delivery refers to adaptations that are intended to tailor a course (or, in some cases, a series of courses) to the individual learner. The intention is to optimise the "fit" between course contents and user characteristics / requirements, *Content Discovery and Assembly* refers to the application of adaptive techniques in the discovery and assembly of learning material / "content" from potentially distributed sources / repositories.

Adaptive Collaboration Support, is intended to capture adaptive support in learning processes that involve communication between multiple persons (and, therefore, social interaction), and, potentially, collaboration towards common objectives.

3 Our Use of Ontology

3.1 Introduction

One of our objectives is to have independent learner model, learning resource and activity. Models used to generate adaptive learning environments in flexible way.

In our work, we have developed ontology to represent models cites above. This ontology has been built by using the language grammar as well as the main concepts related to knowledge.

A possible solution would be to rethink the way of knowledge modeling. The knowledge model, that we should provide, has to be independent of any individual view on the knowledge and should define a shared and consensual terminology. This knowledge model will be used for modeling learning dependencies between resources, describing the actual knowledge state of a particular user and for making inferences on the base of user observations. One possible solution would be to use ontology, which have been developed in Artificial Intelligence to facilitate knowledge sharing and reuse.

3.2 Learner Model

Leaner model is an important component in which help to capture the learner preference, interest, background and / or competence. This model enables the system:

To offer personalized content and guidance, to suggest optimal object observation, to establish users' profiles and surround the actual understanding that they have acquired, to dynamically gather the presented information based on individual needs, to assist specialists so as to offer support in terms of guidance.

In order to build the student model, we have to specify. learners' goals, interests, competencies. We have also to specify how the model to be acquired and maintained.[Paneva 06].We model the user: to give assistance to the learner, to provide feedback, to interpret user behaviour.

On the other hand, such model might be used for teaching system or for other applications. That's why; we need to develop a reusable user model so as to minimize the cost of implementation and to improve the consistency across applications [Kay 99]. With the aim to maximize the reusability and portability of the student model, we have two important standards: the PAPI standard and the IMS LIP standard. The first one stresses on the importance of inter personal relationships. As for the lip standard is based on the classical notion of a resume.

Student modelling researchers adopt different technologies, applications and standards. The earliest ideas of the using ontology for learners modelling have been reported by Chen and Mizoguchi [Chen 99]. Kay [Kay 05] also, agrees about the use of ontology for the reusable and "scrutable" student models.

In fact, the advantages leading to using ontology for the user modelling is that it provides a common understanding of the domain to facilitate reuse and harmonization of different terminologies [Kay 05]. Ontology support reasoning based on inferred rules. Once user characteristics are in ontological representation, the ontology and its relations, conditions and restrictions provide the basis for inferring extra user description. For example, considering a user who wrote a thesis on a subject we can infer that he is a doctor and an expert of the subject. By using ontology and deriving it from the domain ontology , we amplify the possibility that user characteristics will be shared so that systems can use the initialized data for personalization from other systems preventing the user from entering the same information into every system and use the inferred information to adapt the environment.

Razmerita [Razmerita 07] proposes an ontology based user modelling framework which is defines as a generic user ontology describing the different characteristics of a user and the relationships between the different concepts. The proposed user model is extending the IMS LIP which is structured in eleven groupings in including: Identification, Goal, QCL (Qualifications, Certifications and Licenses), Accessibility, Activity, Competence, Interest, Affiliation, Security Key and Relationship. These groupings are implemented as abstract concepts in the user ontology.

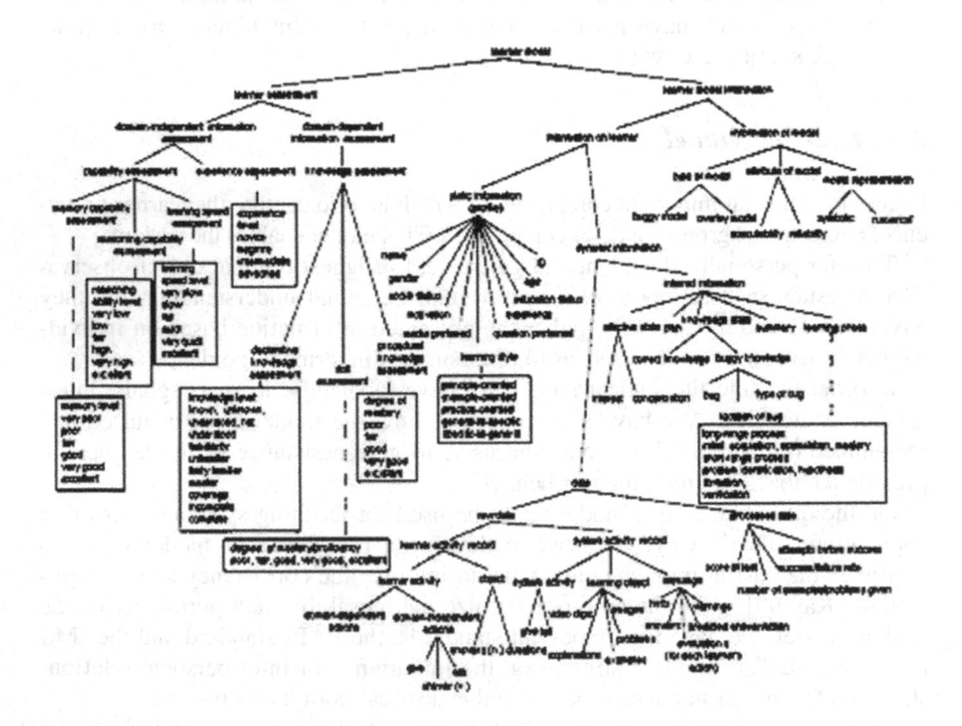

Fig. 1 Learner model ontology inspired from [Razmerita 07]

Ontology based user modelling requires a referential structure which can be static (e.g sub-concept Name of the IMS LIP abstract concept Identification) and an adaptive part which in a learning context need to evolve according to the user's progress in learning, according to his goal, domains of interest which need to be acquired and updated (concepts like Interests, Goal). The dynamic part of the user ontology, such as relationships or activity, could be updated using machine learning techniques [Maedche 03] or other techniques such as trace analysis.

3.3 Learning Resources

In this section, we will take the example of SCORM conformant learning resources in order to explain how to deliver adaptive learning resources according to learner learning style [Drira 06a] [Drira 06b]. SCORM comprises several technical specifications and guidelines for developing learning objects. It was created by the Advanced Distributed Learning (ADL) initiative to meet the Department of Defense (DoD) learning needs in terms of web-based learning contents. SCORM is also an attempt to unify the interests and goals of different groups and organizations that work in the eLearning area.

The specifications that make part of SCORM are organized in two major axes: the Content Aggregation Model, and the Run-time Environment. The Content Aggregation Model provides the specifications for content development, according to the main creation guidelines for learning objects: accessibility, interoperability, reusability and durability. The Run-time Environment defines the mechanisms for establishing the communication between the Learning Management System (LMS) and the learning objects. SCORM is a leading standard for eLearning content development, and a valuable asset for any LMS. In SCORM, the structure of a course is described in a manifest file. A course is a set of hierarchical items (or activities) that reference resources fixed at design time. At run time, all learners are served by the same resources: no consideration of their learning styles is taken into account.

At present it is not possible to use an ontology -based domain model for adaptivity within SCORM. This limits the amount of adaptivity we can export using the SCORM standard. Presently the only way to export ontology -based adaptivity is to strip much of the adaptivity precision. If the details of adaptivity have to be stripped every time the content is to be used on a SCORM-compliant LMS, there seems little point in using ontology to set out adaptivity details in the first place.

An adaptive SCO is an Adaptative Learning Object (ALO). Considering current state of LO (Learning Object) based content adaptation, we suggest that: "ALO is considered adaptive if it is able to be integrated in an adaptation process statically or dynamically. A static participation in an adaptation process is allowed by the appropriate description -metadata- that provides the required information to carry out adaptation. A dynamic participation of a LO is its ability to be modified dynamically (on Run Time) according to dimensions of adaptation related to a particular learning situation."

In order to achieve the adaptive selection respecting to the learner learning style, the SCO has to be adaptive. Unfortunately, the current LOM elements used to describe the SCO do not allow this adaptability [Rumetshofer 03]. Our proposal is based on an ontology for description of SCOs that take into consideration the leaning style supported by a SCO (see figure 2). A reasoner based on a set of rules is also proposed based on the ontology . Its role is to infer the learning styles supported by a SCO (see table1 for examples of rules). It uses a set of rules based on metadata description of a SCO and on Felder Silverman learning style theory [Felder 88]. The considered styles in the reasoner are: Sensing-Intuitive, Visual-Verbal, Active-Reflective and Sequential-Global.

These rules were determined by synthesizing the properties of each style in the Felder theory and by associating these properties to LOM elements. In next table, we present some rules used by the reasonner.

Active learners prefer a LO with higher interactivity level (rule1). The active style supporting level by a LO which interactivity type is active decrease with the decrease of the interactivity level. In other hand, the reflective style supporting level by a LO which interactivity type is active increases with the decrease of the interactivity level (rules 2, 3, 4, 5).

Table 1 Some rules related to Active/Reflective styles.

	Condition	Style and CV
Rule 1	lom.interactivitytype=active et lom.interactivitylevel = very high	Active=1 Reflective=0
Rule 2	lom.interactivitytype=active et lom.interactivitylevel = high	Active =0,9 Reflective=0,1
Rule 3	lom.interactivitytype=active et lom.interactivitylevel = medium	Active =0,8 Reflective=0,2
Rule 4	lom.interactivitytype=active et lom.interactivitylevel = low	Active =0,7 Reflective=0,3
Rule 5	lom.interactivitytype=active et lom.interactivitylevel = very low	Active =0,6 Reflective=0,4

A SCO may fit numerous learning styles but not with the same proportion, so the ontology may contain multiple supported learning styles with their corresponding Confidence Values.

To realize the proposal described in figure 2, a SCO must be conscious of his alternative resources. The LOM relation category allows to define relationships between a LO –learning object- and other LOs. But, the kinds of relationships supported does not allow to express that a LO is an alternative of another one. We suggest that the ontology use the objectproperty **"CanBeSubstitutedby"** to express the possibility of substitution of one LO by another. This allows a SCO to know his alternative SCOs.

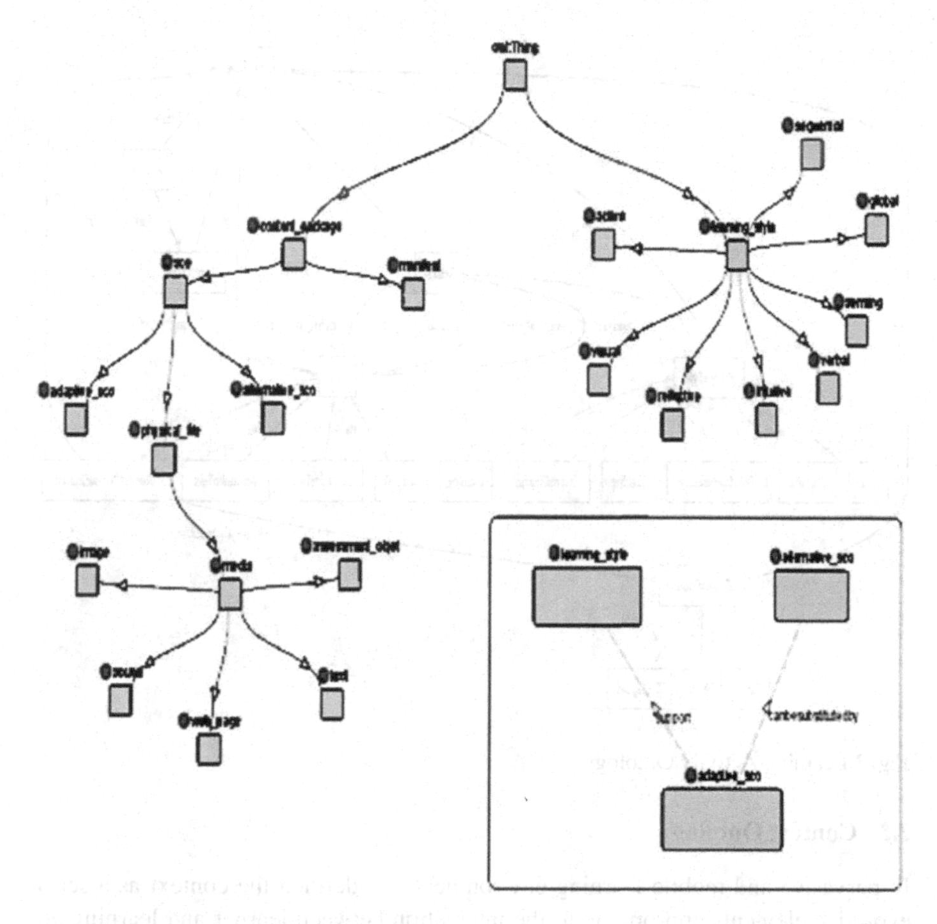

Fig. 2 SCOs ontology.

3.4 Context-aware Adaptive Learning Activities

We identified two ontology: Learning Activity ontology: describes formally components and structure of learning activities; Context ontology : provides means to express formally the knowledge about the learner's and activity's context in which activity is performed;

The activity ontology is inspired from the task model CTT (Concur Task Tree). Indeed, in the ontology of learning activity, an activity consists at first of an "Abstract Task" (see figure 3).The task can be iterative or optional and between two successive tasks, there is a "Relation". Types of relations are the same then in CTT.

An abstract task can consist of various types of tasks: "UserTask", "InteractiveTask" and "ApplicationTask".

The reference of the relevant contextual elements to this activity is done through the use of a concept "relevant context element".

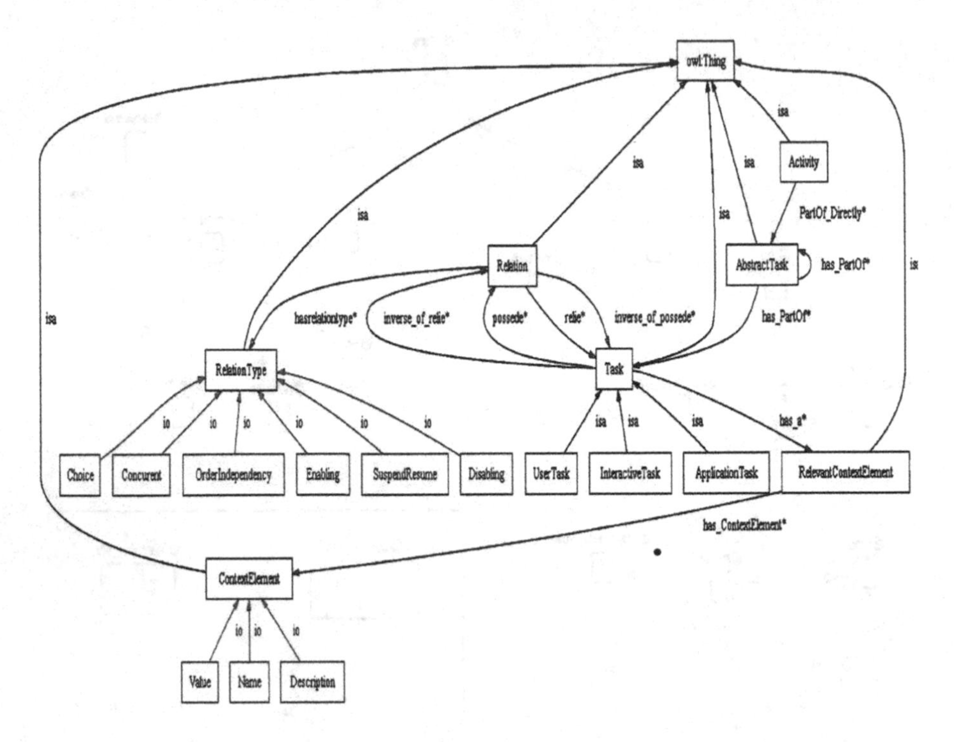

Fig. 3 Learning Activity Ontology.

3.5 Context Ontology

In pervasive and mobile learning environment, we defined the context as a set of evolutive elements appropriate to the interaction between learner and learning application including learner and application themselves. This set of contextual elements composes the context of the interaction that can be divided into two classes: learner context and activity context.

We divide the learner context within mobile and pervasive learning into two major categories: individual context and shared context [Malek 08a]. (figure 4)

To represent the context of learning activities, we based our approach on the activity theory which is a philosophical framework used to conceptualize human activities. [Uden 07] [Kaenampornpan 04] [Malek 09].

In this Ontology , the context consists of "EntityContext". Every EntityContext is formed by an aggregation of contextual elements "ContextElement" that can belong to:

The static type: « StaticContextElement »: does not change during interaction (e.g., season);

The dynamic type: « DynamicContextElement»: changes during the interaction (e.g., noise level). This type specifies the contextual elements that can be adapted

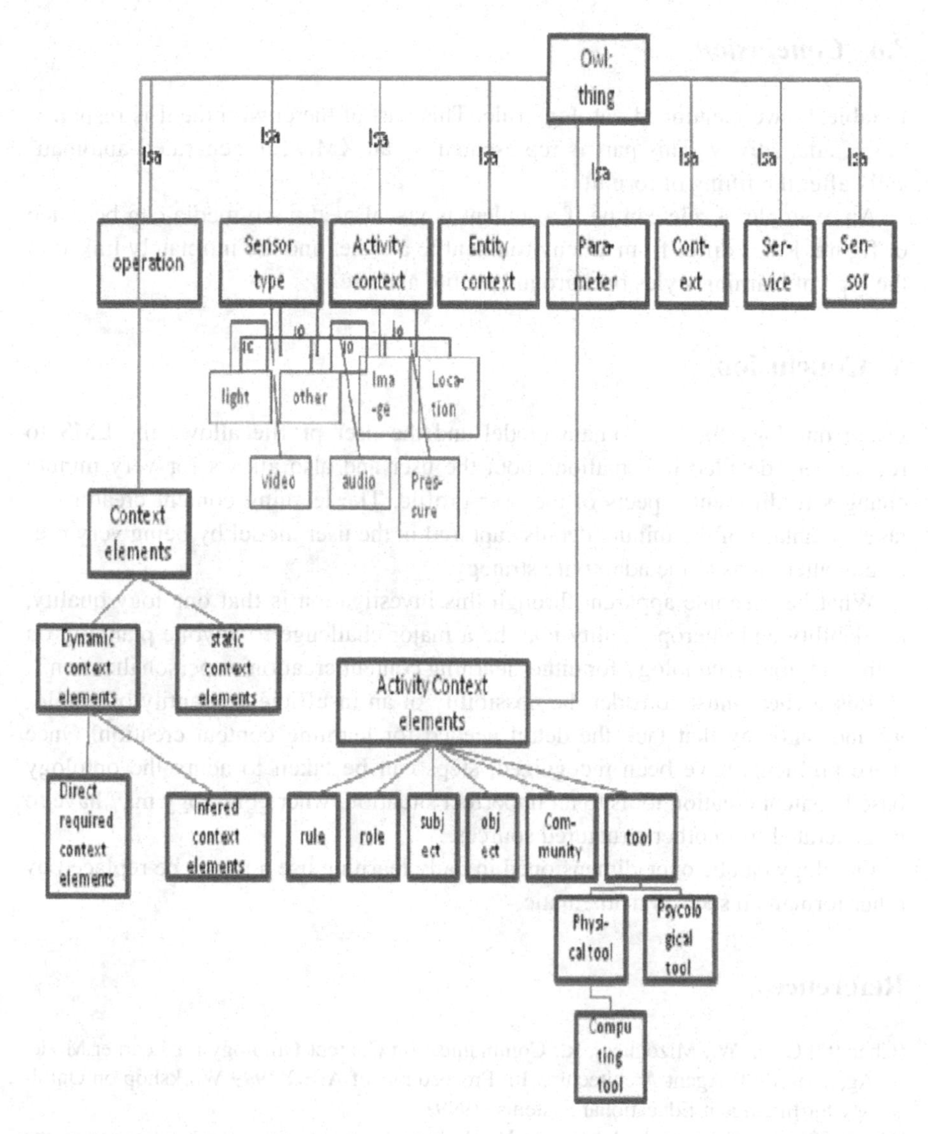

Fig. 4 Context Ontology

by learning activities. A dynamic contextual element can be "DirectlyAcuiredContextElement" through "Services" or "sensors" can have a frequency of update that ensures its newness and can be an "InferedContextElement" may be deducted from other contextual elements by means of" InferenceRules".

3.6 Conclusion

In table 1, we mentioned ontology rule. This part of the environment is responsible of adaptativity. This part is represented by an XML file generated automatically after the filling of formula.

An example of rule can be if a student is visual media sco media can be video or figure. Rules differ from an environment to another and are intimately linked to the type of learning styles background profile and use.

4 Conclusion

Using ontology for the domain model and the user profile allows the LMS to record very detailed information about the user and also allows for very minute changes to different aspects of the user profile. The learning content creator can take advantage of the minute details captured in the user model by being very precise in alterations to the adaptivity strategy.

What has become apparent through this investigation is that ontology quality, availability and interoperability may be a major challenge to anyone planning on using ontology technology for either learning content creation or personalization.

Researchers must consider the possibility of an insufficient quantity of ontology and ontology that lack the detail needed for learning content creation. Once these problems have been recognized, steps can be taken to adapt the ontology based content creation tools to an imperfect situation, where ontology may have to be generated from other structured sources.

Ontology can be over-dimensioned in daily learning use and can be replaced by other formalism such as mathematic.

References

[Chen 99] Chen, W., Mizoguchi, R.: Communication Content Ontology for Learner Model Agent in Multi-Agent Architecture. In: Proceedings of AIED 1999 Workshop on Ontology for Intelligent Educational Systems (1999)

[Chniti 08] Cheniti-Belcadhi, L., Henze, N., Braham, R.: Assessment Personalization on the Semantic Web. Special Issue: Intelligent Systems and Knowledge Management, Journal of Computational Methods in Sciences and Engineering (IO Press) 8(3), 163–182 (2008)

[Drira 06b] Drira, R., Laroussi, M., Derycke, A., Ben Ghezala, H.: Enhancing SCORM to support adaptive and mobile learning content. Wseas Transactions on Advances in Engineering Education 6(3), 571 (2006) ISSN 1790-1979

[Kay 99] Kay, J.: Ontology for reusable and scrutable student model. In: Proceedings of AIED 1999 Workshop on Ontology for Intelligent Educational Systems (1999)

[Laroussi 09] Laroussi, M., Derycke, A.: Extending Lip to provide adaptive mobile learning. International Journal of Interactive Mobile Technologies, iJIM (January 2009)

[Laroussi 10] Laroussi, M.: Pierre-André CARON Adaptativité générique et itérative d'un EIAH aux styles d'interactions des étudiants Implémentation d'un framework de web service pour adapter les fonctionnalités Web 2.0 d'une plate-forme de formation aux styles VAK d'interaction des apprenants, Référence: Conférence EIAH 2011 Mons Belgique (2011)

[Maedche 03] Maedche, A., Motik, B., Stojanovic, L., Studer, R., Volz, R.: Ontology for Enterprise Knowledge Management. IEEE Intelligent Systems 18(2), 26–33 (2003), DOI `http://dx.doi.org/10.1109/MIS.2003.1193654`

[Paramythis 04] Paramythis, A., Loidl-Reisinger, S.: Adaptive Learning Environments and e-Learning Standards. Electronic Journal on e-Learning 2(1), 181–194 (2004)

[Razmerita 07] Razmerita, L.: Ontology -based User Modelling for Knowledge Management Systems. In: Kishore, R., Ramesh, R., Sharman, R. (eds.) Ontology a Handbook of Principles, Concepts and Applications in Information Systems. Springer's Integrated Series in Information Systems, pp. 635–664 (2007) ISBN-10: 0-387-37022-6

[Rumetshofer 03] Rumetshofer, H., Wöß, W.: XML-based Adaptation Framework for Psychological-driven E-learning Systems. Educational Technology & Society 6(4), 18–29 (2003)

Navigating the Educational Cloud

Russell Boyatt and Jane Sinclair

Department of Computer Science, University of Warwick, Coventry, CV4 7AL, UK
{Russell.Boyatt,J.E.Sinclair}@warwick.ac.uk

Abstract. Cloud computing provides a framework which can support many new possibilities for teaching and learning. Cloud-based services and applications are increasingly used by educational establishments to support many aspects of general and educational activity. There is also increasing emphasis on independent learning and open resources. The first part of this paper reviews current published work relating to education and the cloud. One noticeable aspect is that while there is a good deal written about infrastructure, technology and applications there is currently very little on pedagogy. The second part of the paper considers one aspect of this relating to conceptual understanding and navigating resources. We outline our work on a concept-based approach to user-assessment, recommendation and adaptation and consider how this can be used to match resources to an individual's needs.

1 Introduction

The emerging field of cloud computing has shifted focus from local applications and data to a paradigm in which "dynamically scalable and often virtualised resources are provided as a service over the Internet" [13]. The emphasis is on computing as a service, with cloud providers taking care of where and how resources are managed, abstracting these issues from the concern of the end users. Benefits of the approach are considered to include ease of scalability with virtually limitless capacity and elasticity of resources, high availability, increased potential for user mobility and reduction of some routine local IT management issues [14, 36].

Cloud computing can be seen as both driven by and, in its turn, driving technological development [5, 41]. Much of the work carried out so far has been on technologies and architecture [7, 1]. The approach is already being exploited extensively and many providers now offer products which can capitalize on cloud features, encompassing infrastructure, platform and software arenas. The use of cloud in education is a current area of growth seen by many as offering great potential [38]. A recent survey [6] indicates that in the USA 34% of Higher Education institutions and 27% of schools make use of cloud facilities (only the Large Business category has a greater representation than Higher Education with 37%). Many prominent organisations have become involved in cloud provision targeted at education institutions

L. Uden et al. (Eds.): Workshop on LTEC 2012, AISC 173, pp. 179–191.
springerlink.com

(for example, Google [16], IBM [20], HP [19], Microsoft [28]). At a national level, programmes such as that in the UK [21] are underway to provide a private education cloud with appropriate data management, storage and tools.

Published work on education and the cloud has so far been mainly directed at technology and management issues. Aspects such as strategy and governance, and related issues including cloud security and open access are also represented (see for example "The Tower and the Cloud" [24]). Some authors have explored reasons why the affordances of a cloud approach are likely to be beneficial in education. The most common theme is cost, with Sultan [38] referring to the effect of public spending cuts as a likely driver towards cloud solutions in education. Further factors include those noted for cloud use in general: efficiency, reduction in need for in-house expertise and flexibility in resource levels when required [38, 10, 21, 26]. More specific to education are the learning opportunities afforded by accessibility [15]; the ability to reach more students [43] and meeting student expectations [23]. It is also suggested that the shift may motivate a more user-constructed approach to learning and the curriculum rather than the "supply push" of fixed content [37].

Evidence and examples of many universities, schools and consortia using cloud technology are given by a number of authors [38, 12, 31, 42]. Projects in African educational establishments are cited by Sultan [38] who suggests that the benefits of moving to the cloud can be particularly advantageous to developing countries due to considerations of cost, scale and provision of up to date infrastructure. Much institutional cloud use relates to functionality such as email and messaging. Education-targeted applications provide support for many teaching and learning activities (such as hosting learning environments and supporting assessment). However, so far, much less has been done to investigate new ways of teaching and learning via the cloud and finding novel ways in which the technology can be exploited. Little is yet to be found concerning cloud-related pedagogies or evaluating learning. The contributions of the current paper are, firstly, a review of current literature on "cloud education" (sections 2, 3 and 4) and, secondly, a report of the initial stages of our work using conceptual mapping to aid resource recommendation (sections 4 and 5). A prototype implementation in Moodle [30] is also described.

2 Cloud-Mediated Learning

In this section we briefly review existing work on teaching and learning in a cloud context and consider the implications for pedagogy. Cloud computing builds on advances in many areas of computing (such as Web services, Web 2.0, virtualisation, grid computing) and is now being used in educational contexts aligned with these such as mobile learning, distance learning and collaborative learning [35, 22, 15, 40]. There are currently few reports describing specifics rather than relating the general affordances of the cloud to these themes. Commonly addressed themes include social interaction and collaborative working as supported by the increasingly popular Google Apps [16]. For example, Nevin views the common collaborative system gained by the introduction of Google Apps to have "significantly improved the way students and teachers work" [33]. Some authors have started to investigate this type of collaboration in ways which might inform pedagogy. Blau and Caspi [2] investigate students' perceptions on quality and ownership with respect to different modes of collaboration within Google Docs. Kittle and Hicks [27] discuss practical activities which utilize technology to change and improve the collaborative experience of students writing a joint paper.

The collaborative aspect of the cloud is also emphasised by Stevenson and Hedberg [37] who address pedagogical possibilities. They stress [37, p. 322] the need "to investigate the nature in which education might be transformed by the technology and the extent to which the emerging technology-informed pedagogies disrupt traditional teaching and learning". Focusing particularly on Web 2.0, they refer to the lack of research to provide a theoretical framework in which pedagogy can be developed. However, the nature of the cloud means it is not collaborative possibilities alone which must inform pedagogy. For example, mobility is another major feature itself in the infancy of developing conceptual frameworks and establishing effective teaching and learning strategies [34, 39]. The affordances ("anywhere, anytime") and limitations (e.g. connectivity and device limitations) further shape the possibilities for cloud education. Various emergent frameworks (such as [34]) for new modes of collaboration, mobility and other areas cloud education encompasses will, in due course, need to come together to inform pedagogies which exploit its full potential.

3 Learning in Abundance?

One of the positive educational features of the cloud is seen to be the potential for offering more material to a wider range of learners in a greater variety of contexts. Learning is not only taken out of the classroom but can reach constituencies beyond the traditional class. This is not simply a matter of scale and is seen as leading to the "transformative" disruption of education with learner-driven construction of knowledge and socially-constructed, evolving curricula [37, 23]. There is huge potential here as noted by Geith [15, p. 219] who refers to "creating an abundance for teaching and learning". Open Educational Resources (OERs) and Massive Open Online Courses (MOOC) are seen as part of this available wealth. These considerations are already driving government policy in a number of countries as, for example, in the recent UK announcement of the abandonment of the schools' ICT curriculum and suggestion of "an open source curriculum" [17].

Despite the many, exciting possibilities for educational development, a major challenge is how to harness and organise available material to realize the potential. An abundance of learning materials is not the same as abundance in learning. We are now faced with a "superabundance" of information posing the challenge of finding and evaluating what is relevant [25]. There are different aspects to this challenge.

Finding relevant material. At the most basic level there is a need for efficient searching and discovery mechanisms to identify relevant topics.

Educational suitability. A resource is not educationally suitable simply because it addresses a topic of relevance. How to define and provide educational meta-data is an area of research within the OER community. Standards such as Learning-Object Meta-data formalise information that can be understood and processed by learning systems. However, the issues of discovery and suitably are still a major challenge.

Adaptation and personalisation. The power of moving from a "content push" model is diminished if content is still static and available only as "one size fits all".

Learners taking control. To direct their own learning, users must be able to assess their own needs and match these to available materials. Each learner must determine their own prior knowledge and review their understanding as they progress through the material. Success of a learning endeavour may depend to a large extent on this.

These issues are not new: finding suitable material, adapting material for the user and supporting users to guide their own learning are well-known problems. However, the issues gain even greater prominence when dealing with cloud-based learning resources. Any shift from a structured learning environment to learner-driven exploration of knowledge will encounter these issues. In a higher-education environment, a common pedagogy is a series of modules each with a specific focus, structured by prerequisites listing required previous modules. In an environment where learners are free to explore and resources are retrieved from the cloud as required the appropriate pedagogy is not so obvious. In the traditional learning scenario relevant material is likely to have been selected for the learner and the path through this material structured appropriately. In an open-ended learning environment the structure is ephemeral as the learner can be free to choose the resources they wish to use as their understanding and learning objectives evolve.

4 Navigating the Educational Cloud

In this section we consider the issue of navigating the resources available in the cloud to achieve a learning goal. Enabling efficient location of cloud-based resources is a first, most basic requirement. Mechanisms such as keyword-search and tagging are useful tools for discovering new material available for the learner. Given the correct search term or keyword the appropriate resources can be identified. However, this is not necessarily a simple task and finding suitable materials, even from within known repositories is an acknowledged problem [32].

Even supposing that usable and effective search mechanisms to locate relevant material are available, there are limitations to the approach. It requires a certain degree of familiarity with the subject domain – the student must to some extent already have knowledge of the specific terms to search for in a system and be able to recognise the sort of material which will be most helpful to them with their particular background and levels of prior learning. A user-directed model requires the learner to identify suitable resources even if they have a very specific learning objective. Whilst we wish to encourage a learner's ability to reflect and self-assess it may take a significant amount of time to develop this skill particularly in a new topic area. Exploration of a subject is good but unguided exploration is not always desirable or possible for certain groups of students (such as those with learning difficulties), very broad topic areas or where strict time limits are imposed.

As well as knowing how to identify the relevant subject area and find learning materials suitable to the task in hand, the learner needs to take into account their own background knowledge and level of understanding. Again, it can be quite a difficult task to take stock of what one knows and match that to what is needed in order to understand new material. This is an issue already faced in the area of OERs. The challenges include how to capture this information accurately; how to encourage providers of learning materials to include it and how learners can best make use of it. While it is widely acknowledged that self-assessment is a vital part of taking responsibility in learning, it can be a challenging task to do this accurately and one which requires a significant level of educational maturity and awareness [3]. Students who judge wrongly may waste time or become frustrated with material which they discover to be either too elementary or to require large amounts of background work. A student may give up altogether if the difficulties are too great.

Drexler [9] refers to "networked students" who have access to an ever-increasing supply of resources and must demonstrate increased autonomy and self-regulation. In Drexler's case study students work outside the classroom and with access to a wide range of resources but are still learning within a class structure and under teacher supervision. The challenge is how best to balance students' learning autonomy with the structure delivered by the teacher and to move towards greater autonomy as the student becomes more able to direct their own learning. This scenario begins to take into account the wider perspective of open learning but it is still assuming teacher input and direction as necessary. In the broader "anyone, anywhere" view of cloud learning the users may be on their own, so a different model is needed. With no teacher-led direction, mechanisms to support and guide independent learning may make the difference between success and failure for each individual learner.

Using on-line open resources, learners can access (if they are adept at searching) a common baseline of learning material with which they can pursue their learning goal. Strategies and mechanisms are needed which can help learners to identify their current level and match to suitable resources, and also to assist students evaluate and monitor their own learning progress. Ideally, this can be used to help students plan and modify their own learning paths, assisting them in deciding whether they are making good progress or suggesting additional material if remedial help or extra practice necessary. This relates to the agenda of adaptation and personalisation in which different resources and routes can be suggested to individual users depending on a user model. Existing adaptive learning systems are generally within a closed or tightly confederated environment. Work on common platforms and languages such as [8] begins to bridge the gap between different adaptive educational hyper-media systems. However, strategies suitable for adaptation in a general open learning situation would need very flexible student and instructional models in addition to a means of connecting adaptation to the abundant but often poorly documented resources.

Automated pedagogic adaptation is based on assessment of the learner against some abstract model. This could be the user's self-assessment but more accurate information may be obtained by tests which the system can interpret and use to inform adaptation. We have been exploring an approach to mapping student understanding and resource recommendation based on conceptual understanding. This relates to the adaptation approach of cognitive scaffolding [11] in which the system attempts to emulate the personalised dynamic adaptation a teacher would provide in guiding the learner on a path of increasing cognitive difficulty. Central to the "cognitive scaffold" is the idea of cognitive level and the ability to formally represent this and categorize material and test questions accordingly. Again, this is difficult in a more open setting. Further, although this approach can provide adaptation it is the same adaptation for each student in the sense that the scaffold itself must be fixed in advance and, while broad levels of categorization can be supported, differences between individual student's areas of difficulty are less easy to capture.

Our work focuses on identifying key concepts for a subject of study and using this conceptual map as the basis for adaptive assessment and recommendation. It relates to work on concept inventories which have previously been used to develop formative assessment instruments and to assess students' progress through a course of teaching [18]. The emphasis is not on achievement levels but on commonly misunderstood ideas and possible misconceptions which can hinder a student's progress and may remain uncorrected throughout a course of study.

5 Concept-Led Learning

Concept mapping has been used in many areas, most notably in the Force Concept Inventory [18] which covers Newtonian force. Concepts identified are fundamental to a basic understanding of the subject and yet many are often misunderstood. Studies indicate that many students share common misunderstandings which may well persist through a course of study. In such cases, students believe they understand a concept (although their mental construction is wrong) and so do not think it is something they need help with. Evidence that standard teaching and learning activities often leave misconceptions in place indicates that such misunderstandings can be difficult to correct. Further, evidence suggests that students may be able to give correct answers to test questions (and pass the course) even though the basic concepts are misunderstood, indicating rote learning rather than understanding.

These types of misunderstandings have been recognised in areas such as physics and mathematics (where common belief or parlance may be at odds with accepted meaning) but can occur in any area of study. By identifying key concepts and building up a base of known misconceptions, testing can be directed to crucial areas of understanding, quickly exposing problem areas and separating rote learning from understanding. Using concepts as a basis for user modelling and tagging resources with concepts (both prerequisite and those they aim to teach) can provide a framework in which the system directs students to material which is appropriate for their current level of understanding. Further, unlike general topic testing this approach tackles conceptual understanding, uncovers misconceptions and (if possible) directs learners to material that will specifically deal with their own particular needs. Like a human teacher, the approach not only assesses what students do not know but uncovers wrongly-held ideas and addresses them head-on rather than recommending the same, general learning material which the learner may already think they know.

5.1 A New Approach to Conceptual Mapping

Although the concept approach is used in some areas, it is limited by a number of issues. One is that the idea of a "concept inventory" seems to suggest a complete, unequivocal characterisation of a topic which, to develop, requires a large input of expert effort to uncover the "true" concept base. This has proved very difficult to achieve in practice and some efforts have led to much debate between experts as to what should be included. Our work takes a different approach. Working in the area of basic logic and discrete maths we use as a starting point well-documented problem areas and misconceptions together with our own experience of student behaviour. This is brought together to form the basis of a diagnostic instrument and to direct development of online learning resources [4]. The difference is that we do not regard this as an "inventory" but as a starting point to be progressively shaped by student input. Learning resources can be provided for known problem areas. A system which can further analyse students' patterns of answers is used to reveal common errors and those which persist despite study of a unit covering the topic. This agenda, addressed as ongoing action research, has several advantages over an inventory approach. It allows immediate and useful application of what is already known in order to provide support to students rather than waiting to finalise an inventory. Further, it incorporates student-led refinement, allowing identification of further concepts and misconceptions and suggesting where learning material is needed.

A social construction approach fits well with pedagogy based on concepts and misconceptions. Although there are undoubtedly universal misconceptions there also appear to be different patterns of misunderstanding depending perhaps on different backgrounds and learning experiences. It is also likely that the prevalence of some misconceptions may change over time (for example, if a national school curriculum changes, a topic may be added, removed or taught differently resulting in a better or worse overall understanding and different general misunderstandings).

The first step in our action research study has been conducted with international Advanced MSc Computer Science students who need prerequisite knowledge in basic logic and discrete maths in order to pursue advanced level modules. Most have some previous knowledge of the area and 75% judged themselves to have a reasonable understanding (medium or high). An initial concept-based test was administered as the students started the MSc course. Results were analysed to provide information on areas that students did not know and to ascertain general patterns of misconception. Results are also used to provide feedback to students and to direct teaching. The students sat the test again at the end of the first term after taking a module that covered the basics of the topic using general teaching and learning sessions and materials. Further details of the methodology can be found in [4].

5.2 Concept-Driven Technology

A second problem with concept-led approaches so far is use of testing alone. This can provide useful information but limits the approach and does not allow the possibility of using concept information in a broader online education environment. One aim of our work is to investigate how a conceptual framework can be exploited in the context of learning technology, supporting not just testing and feedback but integrating the approach with automated personalised recommendations of learning materials. The advantage of linking to concepts is that variants of learning materials can be provided which address specifically the misunderstandings that each student actually has and which may not be reached by the generic teaching provisions.

Our prototype system uses and extends the Moodle platform [30]. Moodle's mechanisms for extensibility and community-driven plugins allow the core functionality to be enhanced and new activity supported. The prototype introduces a new activity type which supports concepts and misconceptions, allowing these to be associated with test questions and answers using text tags. The test could be self-administered by independent learners. With a class of students we can enforce "before" and "after" testing by incorporating a timing mechanism. Moodle allows concept-based testing to be integrated with general virtual learning environment support such as storing results, giving feedback and managing learning materials.

The second concept-driven activity implemented is to link feedback to concepts, pinpointing for students not simply what questions they have got wrong but, where patterns occur, directly addressing underlying conceptual problems. So far, results have been used to direct module teaching, but the next step is to extend the concept framework to tag learning materials within Moodle and use this to select materials and develop a personalised program of independent study. The learning environment is used to track student progress, provide information and store results for analysis. Patterns of conceptual (mis)understanding are also captured and analysed for each individual student and across the whole cohort. Further extension will allow student tracking against the concept base and adaptation of path depending on progress. For example, our results (in line with previous research) indicate that a common misconception is to interpret logical implication ($p \Rightarrow q$) as if it were conjunction and, further, students may still hold this interpretation after undertaking a course of study. The teaching and learning provision has in this case failed to correct a false belief and presenting the same materials again will only repeat the familiar and fail to reach the real misunderstandings. Our system can incorporate staged variants of learning material, keeping track of the student's learning path up to this point and now suggesting material which specifically tackles the unresolved issues.

A further useful function of the system is for overall analysis of results and statistical investigation of patterns by class and by student.

5.3 Results

We mention here some indicative results from the part our of work addressing propositional logic. Despite students' self-assessment (of having reasonable knowledge) the average initial score on our test instrument of concept-based multiple choice questions was 44.7%. Questions were of a basic nature, covering definitions of connectives and simple deductions. As generally recognised, implication causes most problems and only 16.6% of students answered correctly concerning the basic definition. Biconditional questions were correctly answered by 33.3% of the class. Important from our perspective was the identification of 5 common patterns of misunderstanding for implication (with 2 for disjunction and 1 for biconditional). For example, 20.8% made the "implication as conjunction" error. Initial results for simple deduction showed a 42.9% overall success rate and again, common misunderstandings relating to wrong answers were identified. We took as "common misunderstandings" those made in 3 or more cases. The same concept was tested by several questions to distinguish students making consistent mistakes from those choosing random wrong answers or those making a single error (possibly by accident).

In the repeat test, average scores rose to 63.6% (definitions) and 52.8% (deductions). This shows overall improvement in both areas but the general module teaching is not fully addressing students' needs. Considering the case of implication, students answering the basic definitions questions correctly had risen to 70.8% with 12.5% still interpreting implication as conjunction. Tracking individual students across the two tests shows a variety of profiles. Some students have progressed to answering all questions on a particular concept correctly. Some clearly display a conceptual misunderstanding both before the module and after it. For the former group, it appears the standard module provision has been successful. For the latter group it would be appropriate to recommend material which varies the presentation to address the underlying issue. For other students the pattern is less clear although there are some interesting trends. For example, it appears that some students may in the second test give the correct answer when a question is posed in simplest logical terms but, if the cognitive load is increased by the form or context of the question, they slip back to the same wrong answer they consistently chose in the initial test.

Overall, our results provide further evidence for consistent, common misconceptions which can prove hard to dislodge despite teaching and learning activities. This points to the need to address both what the student does not know and what they incorrectly think they know. Working with a class, diagnostic assessment can inform instructor intervention. For independent learners it is important they have the means and support to identify their own needs and to link that to finding learning materials, amongst the many possibly on offer, which are suitable for their current needs.

6 Conclusions

Cloud technology is being rapidly adopted by educational institutions replacing local infrastructure with cloud services. Take-up is driven by expected financial and administrative benefits, allowing resources to be deployed more efficiently and accessed with greater flexibility by students and teachers. Educational possibilities reported so far are often stated as extensions of web affordances. However, the pedagogical impact of this shift in delivery of learning resources remains unclear. As traditional learning is supplemented or replaced by 'learning in the cloud' we need good strategies for deploying these resources and consider effective pedagogies. An educational institution becomes more than a mechanism of structured learning - it is also required to educate students on how best to navigate the myriad resources on offer. While much has been written about technology, applications and infrastructure relating to the cloud, little attention has been paid to pedagogy.

Different learning situations will require different strategies, ranging from a class with instructor supported by cloud facilities to independent learners working alone in "cloud mode" - that is, bringing together the affordances of mobility, availability and abundance of learning resources using virtualised learning environments as discussed above. On-line courses such as MITx [29] and the various MOOC courses available are demonstrating how on-line resources can be utilised for learning on a large scale. These courses are not seen as full replacements for traditional institutional courses but seek to supplement and enhance existing provision and also reach a much wider audience who would not otherwise have access to such material. Pedagogy is likely to relate to and bring together approaches being developed in the areas which cloud encompasses, such as mobility and virtualisation. However, these are themselves not yet well developed and there is much scope for work in this area. One aspect we have highlighted in this paper is the need to match learner needs with appropriate materials. While there are indeed technological aspects to this we are particularly concerned with students' learning needs and with strategies which can first help identify their conceptual understanding and second relate this to ways to adapt recommendations or personalise learning paths.

The main contribution of our work is to develop a framework in which user modelling based on conceptions and misconceptions allows adaptive recommendation of resources. This supports dynamic and progressive recommendation by the system similar to the way in which a teacher would probe a student's understanding and target misunderstandings which form a barrier to progress. We are currently developing a series of OERs related to common misconceptions identified so far. These will be incorporated into the prototype Moodle system to investigate concept-based strategies for recommendation based on the concept-oriented user model. A further innovation of our work is that, unlike previous "concept inventories" we view a concept base as being organic and user-driven. Further work is needed on how this can be supported in practice in an open environment.

A concept framework could provide the basis for a variety of pedagogic approaches depending, for example, on the level of human instruction associated with the learning situation. Our case study has involved students in a classroom environment, focusing on identifying misconceptions and using this information for resource recommendation. Such an approach could be extended to a more open environment with concept-based user evaluation, matching and adaptation in the absence of tutor-led learning.

Concepts represent one approach to modelling user knowledge and to representing resource prerequisites and content. These activities are themselves just one aspect of learning in an open, resource-rich, user-motivated, collaborative environment. Theories of learning embracing the affordances of the cloud have yet to emerge but the possibilities and challenges are already breaking down the boundaries of the classroom and redrawing the landscape of education.

References

1. Antonopoulos, N., Gillam, L. (eds.): Cloud Computing Principles, Systems and Applications. Springer (2010)
2. Blau, I., Caspi, A.: What type of collaboration helps? In: Proc. of Chais Conference on Instructional Technologies (2009)
3. Boud, D.: Enhancing learning through self assessment. Routledge (1995)
4. Boyatt, R., Sinclair, J.: Supporting concept-based personalisation for masters level CS students. Submitted for Consideration (January 2012)
5. Buyya, R., Broberg, J., Goscinski, A. (eds.): Cloud Computing: Principles and Paradigms. Wiley (2011)
6. CDW-G. The 2011 cloud computing tracking poll (2011), `http://newsroom.cdwg.com/features/feature-05-26-11.html` (accessed January 16, 2012)
7. Chee, B.J.S., Franklin Jr., C.: Cloud computing: technologies and strategies of the ubiquitous data center. CRC Press (2010)
8. Cristea, A., Smits, D., DeBra, P.: Towards a generic adaptive hypermedia platform: a conversion case study. Journal of Digital Information 8(3) (2007)
9. Drexler, W.: The networked student model for construction of personal learning environments: Balancing teacher control and student autonomy. Australasian Journal of Educational Technology 26(3), 369–385 (2010)
10. Educause and Nabuco. Shaping the higher education cloud, `http://net.educause.edu/ir/library/pdf/PUB9009.pdf` (accessed January 16, 2012)
11. Fernandez, G.: Cognitive Scaffolding for a Web-Based Adaptive Learning Environment. In: Zhou, W., Nicholson, P., Corbitt, B., Fong, J. (eds.) ICWL 2003. LNCS, vol. 2783, pp. 12–20. Springer, Heidelberg (2003)
12. Fox, A.: Cloud computing in education (2009), `http://inews.berkeley.edu/articles/Spring2009/cloud-computing` (accessed January 16, 2012)
13. Furht, B.: Cloud computing fundamentals. In: Furht, B., Escalante, A. (eds.) Handbook of Cloud Computing, ch. 1. Springer (2010)
14. Furht, B., Escalante, A. (eds.): Handbook of Cloud Computing. Springer (2010)
15. Geith, C.: Teaching and learning unleashed with web 2.0 and open educational resources. In: Katz, R.N. (ed.) The Tower and the Cloud, pp. 219–226. Educause (2008)
16. Google. Apps for education, `http://www.google.com/apps/intl/en/edu/` (accessed January 16, 2012)

17. Gove, M.: Speech on future of ict in uk schools (2012), http://www.education.gov.uk/inthenews/speeches/a00201868/michael-gove-speech-at-the-bett-show-2012
18. Hestenes, D., Wells, M., Swackhamer, G.: Force concept inventory. The Physics Teacher 3, 141–151 (1992)
19. HP. School Cloud, https://classmate.classlink.us/hp/demo/ (accessed January 16, 2012)
20. IBM. Smartcloud for Education, http://www-03.ibm.com/press/us/en/pressrelease/34642.wss (accessed January 16, 2012)
21. JISC. Cloud services for education and research, http://www.jisc.ac.uk/news/stories/2011/06/cloudservices.aspx (accessed January 16, 2012)
22. Kalagiakos, P., Karampelas, P.: Cloud computing learning. In: 5th Int. Conf. on Application of Information and Communication Technologies, pp. 1–4. IEEE (2011)
23. Katz, R.: The gathering cloud: Is this the end of the middle? In: The Tower and the Cloud, pp. 2–42. Educause (2008)
24. Katz, R. (ed.): The Tower and the Cloud: Higher Education in the Age of Cloud Computing. Educause (2008), http://net.educause.edu/ir/library/pdf/PUB7202.pdf
25. Katz, R., Gandel, P.: The tower, the cloud and posterity. In: The Tower and the Cloud, pp. 172–189. Educause (2008)
26. Katzan, H.: The education value of cloud computing. Contemporary Issues in Education Research 3(7) (2010)
27. Kittle, P., Hicks, T.: Transforming the group paper with collaborative online writing. Pedagogy 9(3) (2009)
28. Microsoft. Microsoft in Education, http://www.microsoft.com/education/en-us/solutions/Pages/cloudcomputing.aspx (accessed January 16, 2012)
29. MITx. MIT Open Learning Initiative, http://mitx.mt.edu/ (accessed February 14, 2012)
30. Moodle, http://moodle.org/
31. Nagel, D.: K-12 budgets begin shift toward cloud (2011), http://thejournal.com/articles/2011/05/26/k12-budgets-begin-shift-toward-cloud.aspx (accessed February 16, 2012)
32. Najjar, J., Klerkx, J., Vuorikari, R., Duval, E.: Finding Appropriate Learning Objects: An Empirical Evaluation. In: Rauber, A., Christodoulakis, S., Tjoa, A.M. (eds.) ECDL 2005. LNCS, vol. 3652, pp. 323–335. Springer, Heidelberg (2005)
33. Nevin, R.: Supporting 21st century learning through google apps. Teacher Librarian 37(2) (2009)
34. Park, Y.: A pedagogical framework for mobile learning. The International Review of Research in Open and Distance Learning 12(2) (2011)
35. Shuai, Q., Ming-Quan, Z.: Cloud computing promotes the progrss of m-learning. In: International Conf. on Uncertainty Reasoning and Knowledge Learning. IEEE (2011)
36. Rittinghouse, J.W., Ransome, J.F.: Cloud Computing: Implementation, Management and Security. CRC Press (2009)
37. Stevenson, M., Hedberg, J.G.: Head in the clouds: a review of current and future potential for cloud-enabled pedagogies. Educational Media International 48(4), 321–333 (2011)

38. Sultan, N.: Cloud computing for education: a new dawn? International Journal of Information Management 30, 109–116 (2009)
39. Taylor, J., Sharples, M., O'Malley, C., Vavoula, G., Waycott, J.: Towards a task model for mobile learning: a dialectical approach. Int. Journal of Learning Technology 2(2) (2006)
40. Thomas, P.Y.: Cloud computing: A potential paradigm for practising the scholarship of teaching and learning. The Electronic Library 29(2), 214–224 (2011)
41. Vouk, M.A.: Cloud computing - issues, research and implementation. Journal of Computing and Information Technology 16(4), 235–246 (2008)
42. Wheeler, B., Waggener, S.: Above-campus services: Shaping the promise of cloud computing for higher education. Educause Review 44(6), 52–67 (2009)
43. Wolsey, K.: Where is the new learning? In: Katz, R.N. (ed.) The Tower and the Cloud, pp. 212–218. Educause (2008)

Constructivist Learning Environment in a Cloud

Jože Rugelj[1], Mojca Ciglarič[1], Andrej Krevl[1], Matjaž Pančur[1], and Andrej Brodnik[1,2]

[1] University of Ljubljana, Slovenia
[2] University of Primorska, Slovenia

Abstract. The paper presents a development of web-based learning environment for constructivist learning in higher education. The main focus in the design was to take into account recent findings of pedagogical research and availability of new technologies in order to create efficient and effective learning support for the engineering students. The central component of the environment is a virtual laboratory, which is defined as a service that can be used in a cloud – LaaS (Laboratory as a Service). The paper also presents our experience with the environment used in Computer Science classes with over 700 students who experienced active forms of learning, collaboration and appropriate feedback.

1 Introduction

Higher education in Slovenia and in many other European countries experienced great changes in the last decade. The percentage of young citizens enrolled into higher education study programs increased dramatically. On the other hand, the amounts of money for higher education in the state budget and consequently number of pedagogical staff and all other resources have not followed this growth. Consequently, efficiency and quality of education have been endangered, while the economic situation is not promising for the near future.

The use of Information and Communication Technologies (ICT) in teaching and learning (***e-learning***) can provide at least part of the solution to these problems. For the purpose of this paper, e-learning is defined as the use of any ICT or applications in the service of learning or learner support. E-learning is important because it can make a significant difference to how learners learn, how quickly they master a skill, how easy it is to study; and, equally important, how much they enjoy learning [1].

There is also a financial impact. Networks and access to online materials offer an alternative to place-based education, reducing the requirement for expensive buildings, and the costs of distance learning materials delivery are low. However, learners still need human support, so the expected financial gains are usually overwhelmed by the investment costs of a new system and the cost of mastering it. Therefore Laurilard [1] claims that it is not possible to build the case for e-learning on cost reduction arguments. It is more reasonable to argue for investment to improve value than to save costs.

L. Uden et al. (Eds.): Workshop on LTEC 2012, AISC 173, pp. 193–204.
springerlink.com

The second catalyst for the interest in e-learning appears to be centred around concern that local higher education institutions might not be able to continue its monopoly on the delivery of education [2]. One area of potential competition is alleged to come from international institutions of higher education. Most activities in e-learning are in the development of courses and their resources and only few institutions have recognized that successful e-learning takes place within a complex system, composed of many inter-related parts. New approaches to learning incorporate and integrate the strengths of face-to-face and online learning in a synergistic manner to create a unique learning experience congruent with the context and intended educational purpose [3]. These approaches are called **blended learning** and combine multiple delivery media. The original use of the phrase "blended learning" was often associated with traditional classroom activities linked to e-learning activities. However, the term has evolved to encompass a much richer set of learning dimensions: blending online and offline learning, blending self-paced and collaborative learning, blending structured and unstructured learning, blending learning and practice [4, 5]. Blended approaches to learning are not just more trendy technology driven ideas that will fade as fast as they come.

In the framework of the SAKE project ("Web Architecture as Educational Technology for Constructivist E-Learning") we have developed architecture for ICT based system for constructivist based learning in higher education. Our activities were organised as a serious of phases, typical for software development: analysis of learning goals and students' needs, the design and development of learning environment and training materials, and the evaluation of the effectiveness of the training intervention. Further, the courses we supported with our architecture include networking, operating systems, and programming. However, in this paper we use a networking ones to describe and evaluate the architecture.

2 Basic Terms

Constructivism [6, 7] is a theory of learning, which claims that students construct knowledge rather than merely receive and store knowledge transmitted by the teacher. In different stages of learning process we use different active forms of pedagogical work in order to engage students to personalize the knowledge [8]. They can adjust the depth of their learning according to their needs and abilities due to open definition of tasks.

Educational assessment is the process of evaluation and documenting, usually in measurable terms, knowledge, skills, attitudes, and beliefs. [9]. Peer assessment is a process where students consider and specify the level, value, or quality of a product or performance of other people in similar situation, usually student within a given class. It represents also an approach to train students how to provide valuable feedback and suggestions for performance improvement. Sluijsmans [10] identified many advantages of peer assessment. Some of the most relevant are that it i) can motivate students and encourage their active involvement in learning, ii) encourages students to become more autonomous in learning, iii) signals students that their experiences are valued and their judgements are respected, and iv) make students to think more deeply, to see how others tackle problems, to pick up points and to learn criticising constructively. Teachers often experience difficulty in evaluation students involved in collaborative activities. The problem lies not only in evaluating the level of learning produced by the process itself, but also in gauging the actual degree to which the individual has actively participated in and contributed to the shared work.

Another important aspect of efficient learning is social dimension of learning. Social constructivism emphasises how meanings and understandings grow out of social encounters [11]. The emphasis is on the learner as an active creator of meanings. Teacher and peer learners enter into a dialogue with the learner, trying to understand the meaning of the material to be learned for each particular learner, and to help learner to refine his understanding. Traditional educational environments are often characterized by a process whereby the teacher assigns a learning activity that is generally carried out autonomously by the student. However, this strips the learning process of a fair amount of its social dimension. So the idea of fostering collaborative learning strategies as a situation in which two or more people learn or attempt to learn something together, presents itself as a means of strengthening this dimension by creating the conditions for individual cognitive development as a result of group interaction [12]. In the case of networked collaborative learning, these strategies are often implemented by assigning a group of students with the task of collaboratively discovering the solution to a given problem or developing a written text based on a given argument. These include different communication and collaboration tools, which are characterized by a variety of unique and powerful information-sharing and collaboration features that offer key advantages, such as allowing learners to be actively involved in their own knowledge construction, as well as improving co-writing processes and facilitating their monitoring [13].

In the next three sections we describe steps undertaken in the process of development of an architecture that supports constructivist based learning. In these steps we use so far described approach and methodology.

3 Analysis of Resources

During analysis of architecture, the designer first develops a clear understanding of the "gaps" between the desired outcomes or behaviours, and the audience's existing knowledge and skills. In the requirements gathering and analysis phase of the system development life cycle we investigated the following factors: students, teaching staff, learning content, institutional framework, and infrastructure.

Our target population are students in engineering and science programs in higher education. Learning goals in majority of courses in these programs are related to higher cognitive taxonomic levels, such as application, analysis, synthesis, and creation of knowledge. Traditional ex-cathedra lectures with textbook, although accompanied by laboratory work, lack in efficiency. Hence, students are supposed to be more actively involved in the learning process, which needs to be transformed into a knowledge creation process. The learning pyramid metaphor is often used to show how the efficiency of learning increases with active participation of students. On average, students after two weeks remember only 10% of what they have read, but they remember up to 90% of what they have done. This claim is well supported in constructivist learning theory and in the methodologies based on this learning theory.

Since our lecturers and teaching assistants are highly qualified for their activities, their role in e-learning environment is different than in traditional lecturing. They are actively engaged in the learning system design and in preparation of learning materials, virtual textbooks, and activities in virtual laboratories. The tutors observe and moderate students' activities in learning management system and in virtual laboratories. They also play an important role in assessment tasks, as these are crucial for efficient feedback in virtual learning environments.

Learning contents are specific for each particular course, but there are some common properties. Previous knowledge of the students, expected learning outcomes, and ascertainment of pedagogical research about motivation and efficient learning methods need to be taken into account. Active forms of learning, collaboration with peers, and construction of new knowledge on previous knowledge are most important.

4 Design and Development of the Learning Environment

The design phase documents specific learning objectives, assessment instruments, assignments, and content. The actual creation of learning materials is completed in the development phase. Within our learning environment, supported by a learning management system, we applied blended learning approach, in order to retain the benefits of face-to-face teaching and class interaction as well as to capture the benefits of virtual learning environment. Innovative learning activities, based on constructivism, were applied in the computer science courses, which were designed as pilot courses as most of participants in the project have background in computer science. We included an option that enables students for new kinds of assessment, such as peer assessment.

Students need opportunities for formative assessment. Feedback helps them develop skills and concepts. As there are usually not enough opportunities to get feedback from staff, peer-learning settings may facilitate additional feedback.

It is a special challenge to implement all these concepts in the ICT supported learning environment. Most of currently available ICT supported learning environments are only a collection of more or less traditional learning materials, some of them originally created and many of them just digitalised textbooks and other learning materials. Students are supposed to read the texts, which are in some cases enriched with illustrations or with some animations and videos. But most of them are based on traditional teaching approach with initial explanation and with some exercises in different forms afterwards. In some cases different multimedia motivational elements are integrated. Therefore we decided to define first a learning design.

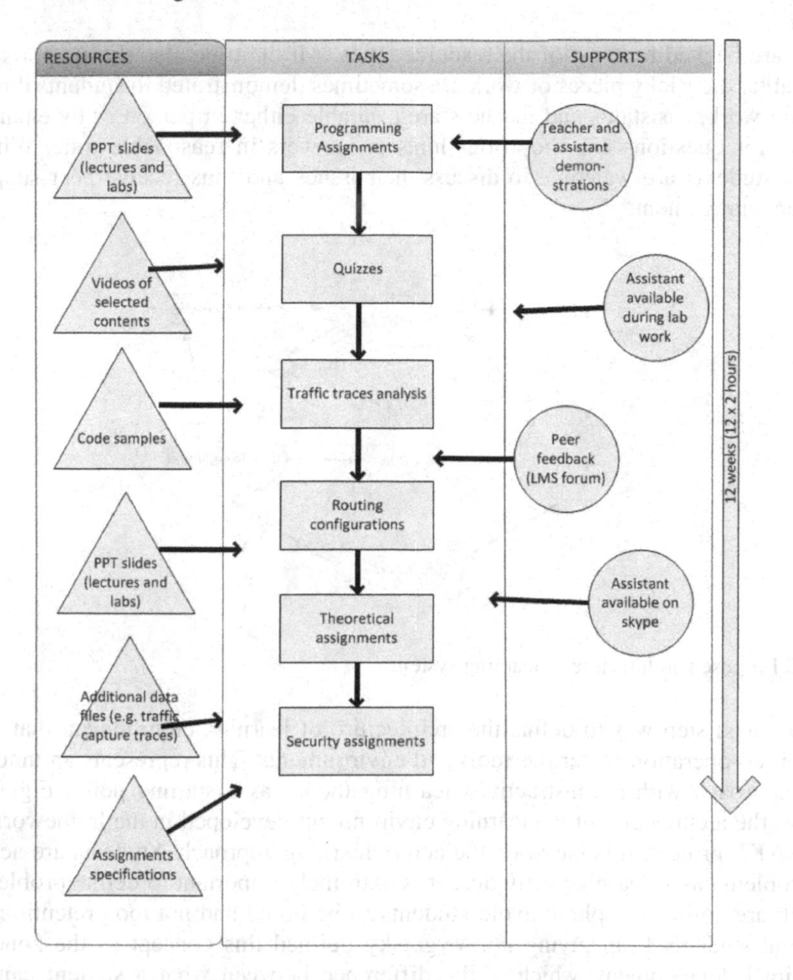

Fig. 1 Learning design of the Computer Communications course

Oliver et al. [14] propose a framework – a structured way to describe learning designs in a unified way, hoping to enable a design to be applied across a variety of learning setting. They also propose a tool, namely diagramming technique they developed to adequately describe the four basic learning designs. By now, our described learning system was used mostly to support rule-based designs and partly also to support strategy-based and incident-based learning designs. Fig. 1 depicts the typical learning design, used in Computer Communications course. The diagram consists of three vertical lanes, the leftmost containing student resources, available in the system. The middle lane contains the tasks students are supposed to accomplish and the right lane shows supports available to students in case of any difficulties during their work.

The tasks are executed in a predefined order, following the contents presented at the lectures. Some of them represent prerequisites needed as inputs into the next task. Resources available to students have different forms, however most of the tasks are backed by most of the resource types. All the time also the supports are available: the tricky pieces of work are sometimes demonstrated to students during the lab work, assistants and teachers are available either in person or by email or Skype for questions and they offer hints or answers in reasonable time. Within LMS, students are welcome to discuss their issues and thus receive peer support and encouragement.

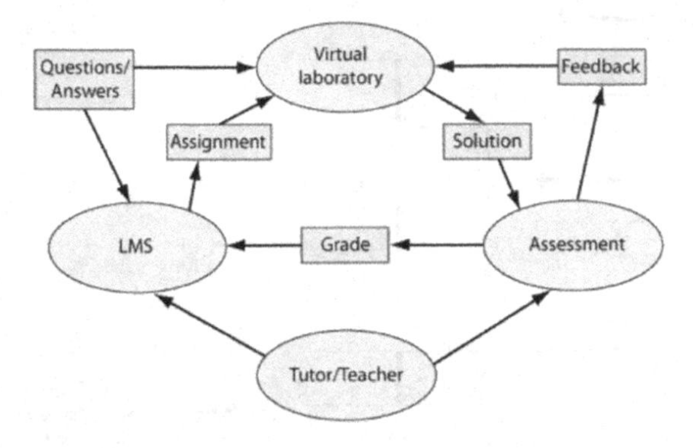

Fig. 2 Proposed architecture of learning system

The next step was to define the architecture of learning environment that will permit co-operation of various tools and environments. This represents an innovative approach with a constructivist learning theory as a starting point. Figure 2 shows the architecture of the learning environment developed in the framework of the SAKE project. It is based on the active learning approach. Students are active in problem based learning activities. It is extremely important to define problems, which are not too simple to avoid students to be bored and not too pretentious to prevent students from giving up. Vygotsky defined this concept as the zone of proximal development, which is the difference between what a student can do without help and what she can do with help. Vygotsky stated that a student follows an example and gradually develops the ability to do certain tasks without help or assistance. The role of education is to provide students with problems, which are in their zone of proximal development and in this way encouraging and advancing their individual learning.

In our architecture, problems are defined in standard learning management system – LMS (e.g. Moodle). After getting familiar with the problem, students are directed into a virtual laboratory, where they can find basic "research infrastructure" with many tools needed to solve the problem. They can also consult a tutor, who supervises activities in virtual lab, and can give different type of advices and recommendations. Learning achievements of self-directed learner depend on learner's self regulated processes. Different categories of Web-based tools (e.g., collaborative and communication tools, content creation and delivery tools, assessment tools, administrative tools) that are part of learning management system, can be used to support different self-regulated learning processes (e.g., goal setting, self-monitoring, use of task strategies, self-evaluating, time planning and management, help-seeking) [15]. Nevertheless, this support is better if the course is facilitated. In e-learning environments tutors are supposed to facilitate planned activities and "take care" of students.

Tutor's main tasks are to create a syllabus that lays out the schedule, requirements and activities of the course, to create code of behaviour within the course, to announce learning goals and expectations, to follow learners' work and monitor their progress, to help learners to progress jointly on the right way, to stimulate a communication among course participants, to actively participate, promote and lead the interactive discussions, to provide answers to questions, feedback and recommendations for activities, and to evaluate and analyze learners' work. Usually the course begins in a classroom, where learners are introduced to their teacher (e.g. instructor, tutor) and other learners in the classroom. Face-to-face classroom sessions take part at appointed times. Schedules for the course should be made on a weekly basis. The schedule determines events (e.g. real-time meetings, videoconferences, chat sessions), readings, (e.g. learning contents that learners must read or view) and activities (e.g. tests, intermediate products for a multi-week project). The activities have deadlines, although learners may complete activities according to their own schedule [16]. This degree of synchronism is important as a motivation and gives students feeling of social inclusion and participation in a group.

An important functionality of our system is an assessment. We implemented facilities that can support both aspects of assessment, formative assessment, which is used to give immediate feedback to students and can be used to direct them in further learning activities, as well as summative assessment at the end of learning activities to measure the achievement of learning goals and to certify the final result of learning [17]. As tutor is not present at all times and students are supposed to do many activities by themselves, formative assessment and its feedback are crucial. Formative as well as summative assessment can be automated and implemented as an expert system or it can be provided by means of "human tutors". The results of formative assessment are presented to students in the virtual lab, while the results of summative assessment need to be sent to LMS.

We mentioned how important the social aspect of learning is within constructivist learning approach. In our architecture at least part of activities are defined in such a way that they stimulate peer learning. To materialize the benefits of peer learning, tutor need to provide 'intellectual scaffolding'. Learning environment needs to support collaborative activities by means of communication support and by means of resource sharing.

5 Implementation of Virtual Computer Laboratory

Virtual Computing Lab (VCL), first developed by North Carolina State University (NCSU), provides a reservation system for virtual and physical resources and enables remote access to these resources. NCSU is using this system to provide applications such as Matlab, Maple, Solidworks, and many others as well as Linux, Solaris operating systems and numerous Windows environments to both students and staff [18], [19]. The VCL software is open source and it is further developed as an Apache Incubator project that is run by the Apache Software Foundation. This enabled us to make the necessary modifications to the project and to integrate into our system as shown in Fig 2.

Our virtual lab needs to support two types of assignments: basic network assignments and advanced infrastructure assignments. Basic network assignments include a single virtual computer where a students needs to complete one or more configuration tasks, e.g. configure a web server, configure an SMTP server, configure a DNS server. Advanced infrastructure assignments span multiple virtual computers interconnected by different network topologies. Students connect network interfaces, configure addressing and routing, and debug connectivity in this type of assignments. While constructive learning and basic networking assignments could have been possible with the existing VCL implementation, support for the advanced infrastructure assignments had to be implemented and integrated into the existing codebase. Although creating a copy of a typical virtual disk should not take more than a couple of minutes on a modern storage system, creating copies of virtual for all the students in a classroom can take several hours. Instead of the default preparation of new virtual machines, in which VCL connects to the storage directly and performs file system operations, we have developed a new approach that utilizes the VMware API and creates a snapshot instead of a complete base image copy. Such a snapshot is created almost instantly and occupies much less storage space since only changes from the base image are stored.

NCSU VCL supported cluster reservations that span multiple virtual computers. However there was no support for advanced networking. We extended the cluster reservation facility to support network topologies and a mixture of different operating systems in a single assignment. Network topology support relies on the usage of several VLANs and a dedicated routing computer that can create VLAN bridges on demand.

Our modifications of the VCL include the integration with a central authentication and authorization infrastructure used at University of Ljubljana. Further, we extended VCL to provide service-based interface becoming a service in a cloud – LaaS, Laboratory as a cloud. Using this interface we integrated VCL with our Moodle LMS, enabling transparent access from LMS to virtual laboratories and transfer of assessment results from VCL to LMS. Moreover, for efficient communication and collaboration in the virtual learning environment we extended the usual forums infrastructure in the LMS by inclusion of fashionable social networks such as Twitter, Facebook, etc. Our pilot virtual lab advanced infrastructure assignment comprised of three VMs (a cluster reservation) with different operations systems, each having three extra unconfigured virtual network interfaces. The isolated virtual network interfaces are bridged together, so that the machines can connect to each other when the students configure them correctly.

6 Research Findings

In the pilot phase, more than 300 students in Computer networking course have used the virtual laboratory. Beside a few small bugs we did not experience technical problems.

The majority of students had two weeks to carry out the assignments, while a smaller group was allowed an extra week. Three different lab assignments were prepared: two of them requiring only one virtual machine and each of them assigned to approximately half of the students, while the third required three virtual machines and was assigned to all students. If the two assignments were completed each in the first try, a total of four reservations were needed, however the students were allowed to enter the virtual lab, make any number of reservations and use the virtual machines at will. The reservations were set to 60 minutes at a time, but the students were able to prolong them if they weren't finished in time.

Our cloud consists of eleven quad-core physical servers, each with 30 slots - for a maximum of 30 concurrent virtual machines. In addition to these core "compute" servers, our cloud also contains two management node servers (the first for VCL management tasks, the second for hypervisor management and support), a dedicated VCL network router/bridge, (used for bridging isolated virtual network interfaces between virtual machines) and a dedicated data storage server (used for storing virtual machine images). In the observed 20-days period the physical servers have hosted from 167 to 425 virtual machines, with an average of 258, while each slot hosted 10 virtual machines on average. A total of 317 students have made at least one reservation, while the total number of virtual machine (VM) reservations in the observed 20 days was 2840. Note that the third assignment takes three virtual machines and the actual number of assignment reservations was 1347. In the rest of this section, the term reservation refers to a virtual machine reservation unless stated otherwise. Students spent a total of 934 hours in virtual lab, of those 550 hours in the third assignment. The average reservation duration was 44 minutes and 24 seconds; more detailed frequency diagram is shown in Figure 3.

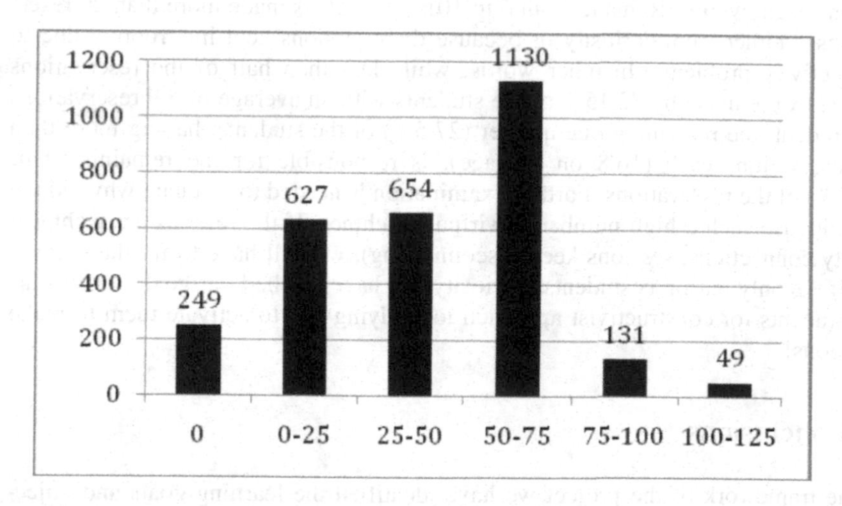

Fig. 3 Number of VM reservations by reservation duration

The reservations with short duration (0-25 minutes) represent student testing - how the system works and how the assignments look like, since we estimate it is impossible to complete the assignments in such short time. The majority of students were able to complete the assignments within 75 minutes, while the longest reservation time was 120 minutes.

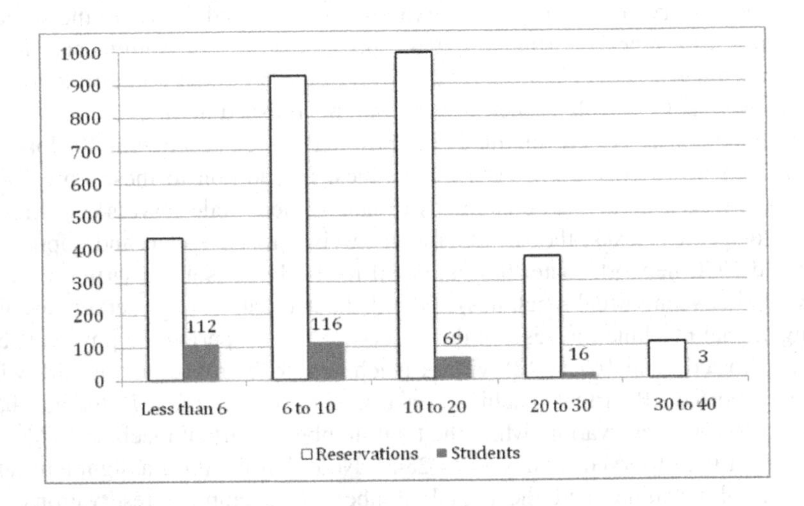

Fig. 4 Number of reservations belonging to the same student.

Figure 4 shows the frequency break-up of the number of virtual machine reservations made by the same student. The majority of students needed less than 10 reservations to accomplish the assignments (112 students made less than 6 reservations each, while 116 made from 6 to 10). 19 students made more than 20 reservations – either from curiosity or because their sessions kept interrupting due to connectivity problems. In other words, while less than half of the reservations (48.9%) were made by 72.15% of the students with an average of 5.9 reservations per student, the remaining one quarter (27.5%) of the students, having more than 10 reservations each (16.8 on average), is responsible for the remaining half (52.1%) of the reservations. Further examination is needed to evaluate why did the students use such a high number of virtual machines. If the reasons are technical (faulty connections, sessions keep disconnecting), we will have to fix the system. But if the only reason is students' curiosity, we have reached our goal – to encourage students for constructivist approach to studying and to activate them to make decisions!

7 Conclusions

In the framework of the project we have identified the learning goals and objectives, the student needs, and other relevant characteristics about students and teaching staff to design an open learning architecture. The main requirements for the learning environment and other infrastructure needed to implement the selected courses were also considered. In the design phase we have defined the architecture of the system and outlined the main elements of learning activities and contents for the selected courses, based on the specified learning objectives and goals. The actual creation of the content and learning materials as well as the definition of activities in virtual labs were carried out in the development phase. Virtual labs and learning management system were implemented as a service in the cloud – LaaS. We have found out several interesting facts about the dynamics of virtual lab use and about the workload patterns in the virtual lab in the evaluation phase made possible due to time-and-space shift paradigm. Qualitative evaluation of feedback from students immediately after the course showed that students found the assignments difficult and that the new approach to learning was confusing for some of them. However, the survey made after the examination period showed the feedback was more positive. The overall perception of the importance of computer communications area was higher as it was at the beginning of the course.

References

[1] Laurillard, D.: E-Learning in Higher Education. In: Ashwin, P. (ed.) Changing Higher Education: The Development of Learning and Teaching. Routledge (2005)

[2] Alexander, S.: E-learning developments and experiences. Education + Training 43(4/5), 240–248 (2001)

[3] Garrison, D.R., Vaughan, N.D.: Blended Learning in Higher Education. Jossey-Bass, San Francisco (2008)

[4] Singh, H.: Building Effective Blended Learning Programs. Education Technology 43(6), 51–54 (2003)

[5] Lapuh, J., Rugelj, J.: Blended learning -an opportunity to take the best of both world. Int. Journal on Emerging Technologies in Learning 2(3), 71–75 (2007)

[6] Nančovska Šerbec, I., Strnad, M., Rugelj, J.: Students' attitude to active forms of e-learning. In: Činčin-Šain, M. (ed.) Rijeka: Proc. MIPRO 2009, pp. 100–103 (2009)

[7] Ben-Ari, M.: Constructivism in Computer Science Education. Jl. of Computers in Mathematics and Science Teaching 20(1), 45–73 (2001)

[8] Nančovska Šerbec, I., Strnad, M., Rugelj, J.: Active learning in computer science courses in higher education. In: Kinshuk (ed.) Proceedings of CELDA 2009, pp. 538–540. International Association for Development of the Information Society, Roma (2009)

[9] Nančovska Šerbec, I., Strnad, M., Rugelj, J.: Assessment of active forms of learning in the higher education. In: Auer, M.E. (ed.) ICL 2009. Wien: International Association of Online Engineering, pp. 489–496. University Press, Kassel (2009)

[10] Sluijsmans, D., Dochy, F., Moerkerke, G.: Creating a learning environment by using self-, peer- and co-assessment. Learning Environments Research 1, 293–319 (1999)

[11] Hansen, T., Dirckinck-Holmfeld, L., Lewis, R., Rugelj, J.: Using telematics for collaborative knowledge construction. In: Dillenbourg, P. (ed.) Collaborative Learning: Cognitive and Computational Approaches. Elsevier Science, Pergamon (1998)

[12] Lippman, P.: Evidence-Based Design of Elementary and Secondary Schools. Course Smart Series. John Wiley & Sons (2010)

[13] Trentin, G.: Using a wiki to evaluate individual contribution to a collaborative learning project. Journal of Computer Assisted Learning 25, 43–55 (2009)
[14] Oliver, R., et al.: Describing ICT based learning designs that promote quality learning out-comes. In: Beetham, P., Sharpe, R. (eds.) Rethinking Pedagogy for a Digital Age. Routledge (2007)
[15] Dabbagh, N., Kitsantas, A.: Using web-based pedagogical tools as scaffolds for self-regulated learning. Instructional Science 33(5-6), 513–540 (2005)
[16] Horton, S.: Web Teaching Guide: A Practical Approach to Creating Course Web Sites. Yale University Press, New Haven (2000)
[17] Lapuh Bele, J., Rugelj, J.: Providing feedback in web-based learning. In: Auer, M.E. (ed.) 10th International Conference, ICL 2007. ePortfolio and quality in e-learning. Internat. Assoc. of Online Engineering, Wien (2007)
[18] Vouk, M.A.: Cloud Computing – Issues, Research and Implementations. Journal of Computing and Information Technology 16(4), 235–246 (2008)
[19] Schaffer, H.E., Averitt, S.A., Hoit, M.I., Peeler, A., Sills, E.D., Vouk, M.A.: NCSU's Virtual Computing Lab: A Cloud Computing Solution. Computer 42(7), 94–97 (2009)

Learning Strategy Recommendation Agent

Zhendong Niu, Peipei Gu, Wenshi Zhang, and Wei Chen

School of Computer Science and Technology, Beijing Institute of Technology, Beijing, 100081, China

Abstract. Nowadays, it has been an important issue to adaptively recommend learning strategies for every learner in intelligent tutoring systems (ITS) that covers various areas and subjects. In this paper, three models for learners, learning strategies and learning strategy-oriented services are proposed. The C4.5 decision tree algorithm is adopted to construct a learning strategy tree which contain popular learning strategies used in ITS. Based on those models and the learning strategy decision tree, a learning strategy recommendation agent is proposed in our learning strategy recommendation system (BIT-LSS) to adaptively recommend learning strategies for learners. Questionnaire surveys and experiments are conducted to demonstrate the efficiency of the learning strategy recommendation agent in BIT-LSS.

Keywords: Intelligent tutoring system, strategy recommendation agent, learning strategy, learning profile.

1 Background and Motivations

With the rapid development of information technology, education technology for online learning is adopted by many colleges and universities [1]. Intelligent tutoring systems (ITS) are becoming an inevitable trend of education reform throughout the world. It is widely believed in tutoring researches that tutors are adaptive to the needs of the tutees [2]. Learners in ITS systems have different interests, previous background of knowledge and acquire information in different ways [4]. A learning strategy can be effective for one learner and not for another. One of the most important issues in ITS systems is how to choose learning strategies adaptively according to tutoring learners' personal needs. Learning strategies can be defined in many ways, but the general idea is that learning strategies are behaviors and thoughts that a learner engages in during learning process [3].

The ITS systems are often evaluated by their ability of making tutoring decisions for a learner [5] [6]. Appropriate use of learning strategies can contribute to success and efficiency of learners learning. Many researches have been conducted in language and mathematics learning areas. Some researchers add specific learning strategies to training systems to help learners to master knowledge. Lulis *et al.* apply analogies to the tutoring strategies available by using natural language generation techniques and by modeling the behavior of experts in the ITS system

L. Uden et al. (Eds.): Workshop on LTEC 2012, AISC 173, pp. 205–216.
springerlink.com

which they have built for cardiovascular physiology [7]. Hashemi observes the language learning strategies in the Iranian EFL learners and the impact of the gender factor in the process of language learning [8]. Arroyo *et al.* have built a Math Facts Retrieval Training (MFRT) system, and experiments demonstrate that MFRT can effectively improve learners' performance [9]. Some researches have been carried out to analyze the relationship between different learners and learning strategies. Rubin proposes that teaching good language learning strategies to learners whose scores are low might be useful for improving their performance [10]. Besides, some studies are concentrating on learning strategy recommendation. Bull constructs a learner model which contains representations of learning styles and the current strategy usage, and builds a system called LS-LS to infer potential interesting learning strategies for an individual to help him/her to become an effective learner [6].

In recent years, technologies in ITS systems have developed very fast. More and more ITS systems cover rich subject themes, different resource types, and different areas. How to use learning strategies adaptively to help learners to become more effective is an important problem.

Most of the previous researches concentrate on using learning strategies in specific learning area, such as math or language. Few work concentrate on learning strategies of general application in ITS systems which is composed of resources cover various kinds of subjects in different areas.

In this paper, based on our previous work [11], we propose three models to formulate learners, learning strategies and learning strategy-oriented services. We construct a learning strategy tree by using C4.5 classification algorithm. A learning strategy recommendation agent is presented to help learners choose appropriate learning strategies. Questionnaire surveys and experiments are conducted to demonstrate the efficiency of using the learning strategy tree in BIT-LSS.

The rest of the paper is organized as follows. Sect. 2 presents three models that formulate learners, learning strategies and learning strategy-oriented services. Sect. 3 presents our learning strategy recommendation process. Sect. 4 is the experiments and evaluation. Sect. 5 draws the conclusion of this paper and presents the future work.

2 Models of BIT-LSS

In the reality, there are various factors that influence learners to master knowledge and apply learning strategies: attribute set of learners themselves, such as static properties, dynamic properties, and emotion factors; knowledge structure; teachers' factors, such as experience, instruction methods, and feedback. For the sake of constructing adaptive learning strategy recommendation agent, we should model learners, learning strategies and learning strategy-oriented services.

2.1 Learner Model

As Fig. 1 shows, BIT-LSS learner model provides information about learners' static properties, dynamic properties, affective information, history of learning strategy choosing, and test results of all tests in BIT-LSS.

Static properties and extension contain personal information, such as age, education, location, sex, etc. These might have no direct access to learning process in learning system, but these can be factors influencing learning strategy choosing.

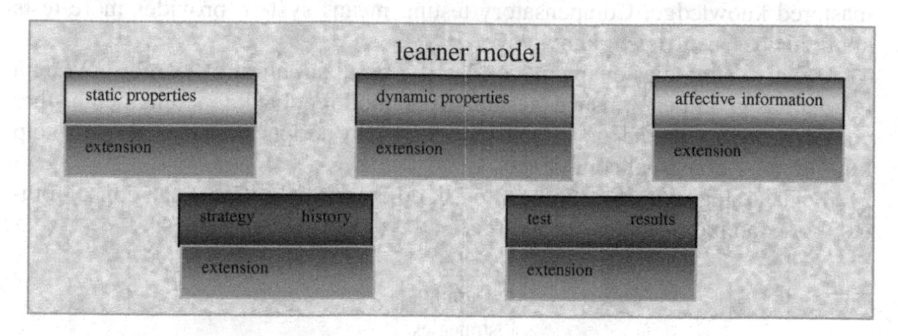

Fig. 1 Learner model in BIT-LSS

Dynamic properties and extension describe learners' active learning tracks, learning styles, and interactive method.

Affective information can help us to understand learners' mood information. We can reward a learner when he/her get good performance, and try to use positive methods to help a learner re-construct his/her confidence when he/she gets bad test results.

History of learning strategy choosing can help us to get learners' learning strategy selection preference, and find the effects that different strategies achieve.

Test results are one hard index in BIT-LSS. They support evaluating a learner by means of giving him/her tests which cover knowledge information that he/she wants to learn and master in BIT-LSS.

2.2 Learning Strategy Model

Learning strategies have been defined by Scarcella & Oxford as "specific actions, behaviors, steps, or techniques – such as seeking out conventional partners, or giving oneself encouragement to tackle a difficult language task – used by learners to enhance their own learning" [12]. Oxford has identified six major groups of learning strategies: cognitive, metacognitive, memory-related, compensatory, affective, and social [13].

Oxford concentrates on study leaning strategies usage in language learning area. In order to construct general learning strategy model for BIT-LSS that has various kinds of learning resources, the learning strategy model representation in BIT-LSS is shown in Fig. 2.

The five groups of learning strategies in BIT-LSS are as follows:

1) Metacognitive strategies help learners to manage the learning process in general, e.g., learners make plans for whole learning goal and summarizes learning outcome.
2) Memory-related strategies enable learners to learn or retrieve resources and knowledge which have been learned earlier.
3) Compensatory strategies help learners learn and master missing knowledge. Compensatory teaching can help learners to learn more about his/her non-mastered knowledge. Compensatory testing means system provides more tests about his/her insufficient knowledge.
4) Affective strategies enable us to identify mood situation of learners. When a learner has good performance, system gives good feedback to strengthen his/her confidence. If a learner is in bad mood, system adopts positive ways to help him/her to re-construct confidence.
5) Social strategies enable learners to ask others for help. Users can consult others. For example, learners can reference others' learning strategies.

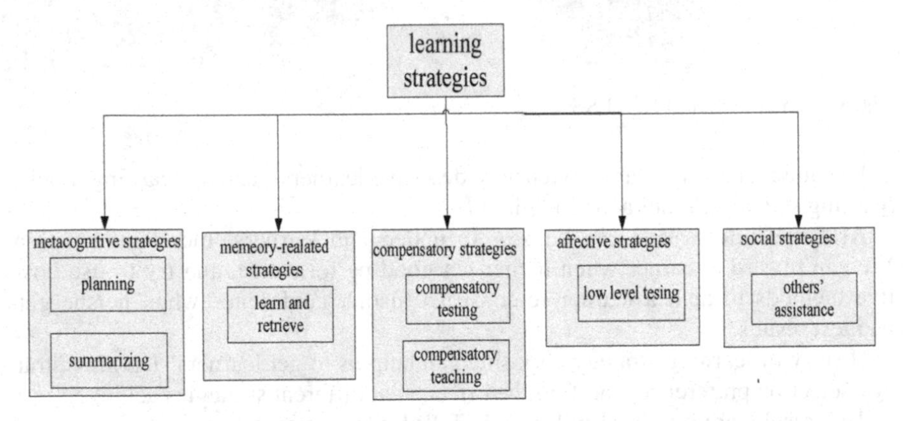

Fig. 2 Learning strategy model representation

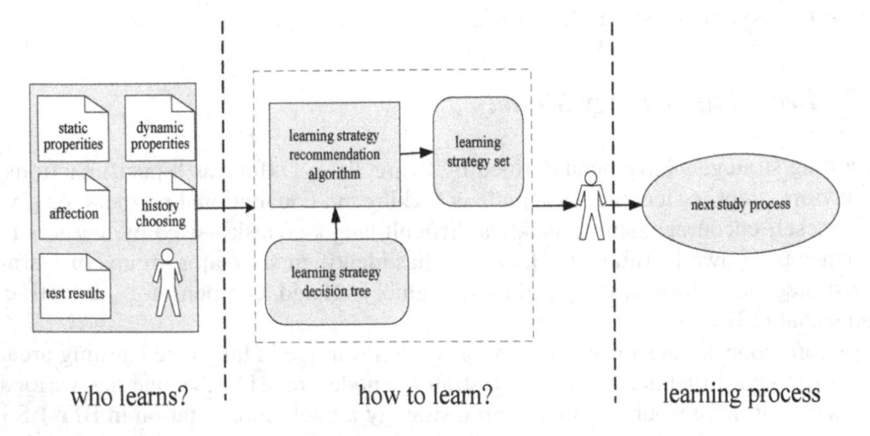

Fig. 3 Learning strategy-oriented service model

2.3 Learning Strategy-Oriented Service Model

BIT-LSS provides adaptive learning strategies for learners. It is a learning strategy-oriented service system.

Learning strategy-oriented service model can be divided into three parts: 1) who learns? 2) how to learn? 3) learning operations. As Fig.3 shows, in the first part, we need to know who the learner is and get his/her information. In the second part, according to the learning strategy decision tree we made (we will show more tree construction information in the next section) and his/her information, learning strategy recommendation algorithm can provide a learning strategy set for him/her to provide learning reference about how to learn. He/she can make adjustment on learning strategy recommendation and decide what to do in the next study process.

3 Learning Strategy Recommendation Process

The learning strategy recommendation process can be divided into two parts: the decision tree construction processing, and strategy recommendation agent.

3.1 Decision Tree Construction Processing

Classification is a hotspot in the research of data mining. The most widely used classification model is decision tree algorithm [14]. Decision tree algorithm is applied in the data mining of learning strategy samples offered by the questionnaire classification. We construct a learning strategy classifier based on C4.5 which is proposed by Quinlan [15].

C4.5 uses information gain ratio as a heuristic function to extend attributes. Information gain ratio equals ratio of information gain to split information.

The categories are U_1, U_2..., U_m. Training set is S. Attributes are A_1, A_2... A_p . S is assumed that be divided by attribute $A_t = \{A_{t1}, A_{t2}... A_{tn}\}$ $(1 \leq t \leq p)$ to be n sub-sets S_{t1}, S_{t2}... S_{tn}. S_{tj} $(1 \leq j \leq n)$ is composed of samples which attribute A_t of them equals A_{tj} in S. $|U_i|$ is the quantity of samples which belongs to category U_i in S. $|U_i|/|S_{tj}|$ is the conditional probability $P(U_i|A_{tj})$ when attribute A_t of samples equals A_{tj} and category of samples is U_i.

The definition of total information gain ratio $GainRatio(A,S)$ when S is divided by attribute A is as follows:

$$GainRatio\,(A_t, S) = \frac{Gain\,(A_t, S)}{SplitInfor\,mation\,(A_t, S)} (1 \leq t \leq p) \quad (3.1)$$

where

$$SplitInfomation(A_t, S) = -\sum_{j=1}^{n} \frac{|S_{tj}|}{|S|} \times \log_2 \frac{|S_{tj}|}{|S|} \quad , \tag{3.2}$$

and

$$Gain\ (A_t, S) = -\sum_{i=1}^{m} \frac{|U_i|}{|S|} \times \log_2 \frac{|U_i|}{|S|} - \sum_{j=1}^{n} \frac{|S_{tj}|}{|S|} \times \sum_{i=1}^{m} \frac{|U_i|}{|S_{tj}|} \log_2 \frac{|S_{tj}|}{|U_i|} \quad . \tag{3.3}$$

The procedure of construction for decision tree is as follows:

Input: training set *S,* attributes are $A_1, A_2 \ldots A_p$ and categories $U_1, U_2 \ldots, U_m$.
a) If termination criterion is met or set is indivisible, mark it as a leaf node.
b) Go to step *i)* if all the training sub-sets is marked.
c) For every attribute $A_{t,}$ $(1 \leq t \leq p)$, go to *d)~ f)*.
d) Calculate information gain ratio of attributes according to *(3.1)*, *(3.2)*, and *(3.3)*.
f) Choose best gain ratio of attributes $A_{t,}$ $(1 \leq t \leq p)$, and divide training set *S into sub-sets* $S_{t1}, S_{t2} \ldots S_{tn}$. S_{tj} $(1 \leq j \leq n)$ according to $A_{t,}$ $(1 \leq t \leq p)$.
g) Repeat step *a)~ f)* for new sub-sets $S_{t1}, S_{t2} \ldots S_{tn}$. S_{tj} $(1 \leq j \leq n)$.
h) Output the final decision tree of learning strategies.

However some attribute values in sample might be empty, C4.5 algorithm assigns one weight value for every sample. At first time, all weight value is *1.0*. When attribute *A* divides training set *T* into *S* sub-sets $T_1, T_2, \ldots, T_s$, T_i contains all sample. *A* equals a_i and *A* equals *null* in training set *T*. The weight value of T_i is proportional to ratio of sample number of T_i which *A* isn't *null* and sample number of *T* which *A* isn't *null.*

C4.5 algorithm estimates classification error based on the training set. It might make the decision tree inclined to overfitting. Therefore, C4.5 algorithm adopts pessimistic pruning method.

3.2 Learning Strategy Recommendation Agent

The proposed models which provide learners' information and learning strategies are designed to simulate the operation of a strategy recommendation agent. BIT-LSS is targeted at three kinds of people: 1) novice. They have no related knowledge background. Their learning behaviors in learning system are random. Or they don't use system before. 2) Participants which use system for some time, but they have not adopted learning strategies. 3) Participants which have learning strategy selection history. They choose learning strategies themselves or recommended by the agent.

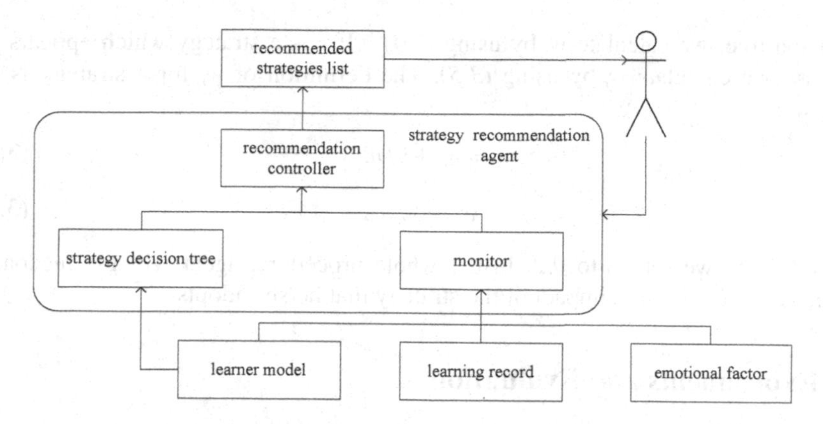

Fig. 4 Learning strategy recommendation agent

Strategy recommendation agent presents procedure of learning strategy recommended as shown in Fig. 4. The agent includes three parts: 1) monitor, 2) the decision tree constructed in Sect. 3.1, 3) recommendation controller. At first, monitor pays close attention to learner model which contains adopted strategy history of learners, learners' learning records and learners' current emotional state. When a learner asks the agent to recommend the next learning strategy, learning strategy recommendation controller recommends appropriate strategy list for him/her according to the learning strategy decision tree which has been constructed in Sect. 3.1 and learner comprehensive information which monitor gets from learner model, learner learning records and learner's current emotional state. If a learner picks one strategy from agent recommendation list as his/her next operation strategy, the strategy would not change until he/she asks the strategy recommendation agent to recommend one next time or he/she left this procedure.

The rules that we choose to select the strategy in strategy recommendation controller are as follows: 1) if a learner is a novice, the agent selects the most popular strategy in the system for him/her. 2) If a learner has some learning behaviors, the agent selects strategies based on the strategy decision tree constructed in Sect. 3.1. 3) If a learner has strategy recommendation history, every strategy in history has a recommendation weight value which is used for calculation in *(3.4)* and *(3.5)*. The agent recommends the best appropriate strategies by calculating recommendation weight values of strategy selected by the strategy decision tree and strategies in learners' history choice.

In this paper, we incorporate w_r which plays a balance in strategy history and strategy from the decision tree as recommendation weight value of a candidate strategy *r*. The candidate strategy set contains strategy recommended by the strategy decision tree and strategies in adopted history. *m* is the size of candidate strategy set. *s* is the frequency of this strategy which appears in history. w_O is the weight value for strategy selected by the strategy decision tree. w_O is a fixed value which is assigned by a tutor. If *r* is a strategy which is recommended by the decision tree, we calculate w_r by using *(3.4)*. If r is a strategy which appears in history, we calculate w_r by using *(3.5)*. The definition of w_r for a strategy is as follows:

$$w_r = w_0 + s / m \ , \tag{3.4}$$

$$w_r = s / m \, . \tag{3.5}$$

In BIT-LSS, we set w_0 to *0.2*. In the whole procedure, agent will get feedback from learners to see the impact of the strategy that he/she adopts.

4 Experiments and Evaluation

To evaluate the efficiency of the proposed strategy recommendation agent and learner model, a series of questionnaire and experiments are conducted on students in Beijing Institute of Technology who are arts and science students.

4.1. Learners' Data Processing

There are various learning behaviors in BIT-LSS. A learner can learn from video, ppt, or word. He/she could communicate with other learners in the system. He/she could do online tests or submit offline works.

We make a questionnaire that collects *30* questions, involving learners' attributes and learning strategy classification, to provide a relationship between learners' attributes and learning strategies. Though there are 27 item choice questions and 3 questions to fill in the blank in the questionnaire, data from the questionnaire which are static information might have varying degrees of difference in learners' actual operations. To some extent, results from the questionnaire could reflect learners' learning strategies choosing inclination. Most of participants are master and doctor who major in arts and science in college. According to some surveys, we tend to believe that master and doctor students have more mature opinions about learning procedure and learning suggestion. This work began at Jun 2011 and ended at Oct 2011. We release about *120* copies and receive *97* effective response copies back.

For constructing a learning strategy tree, we have a procedure of data preprocessing. We assume difficulty degrees of user learning preference are *1*, *2*, and *3*, which represent low, medium and high, respectively. Personality characteristic values describe proportion of positive factor in personality characteristic of learners. Learning ways can be divided into *1*, *2* and *3*, representing video, ppt and others. In the learning affection, we use *1*, *2* and *3*. to represent low, medium and high learning affection. Test results describe test scores that learners achieve at last. *1* indicates learners get scores at *0~60*. *2* indicates learners get scores at *60~80*. *3* indicates learners get scores at *80~100*. Learning strategy choice is the best choice satisfied current attribute sets. For example, A represents compensatory strategy; B represents others-help strategy etc. Table 1 is a sample of training data after reorganization.

Table 1 Data after reorganization

Personality characteristic	Difficulty degree	Learning way	Affection	Test results	Strategy choice
1	2	1	1	1	1
0.75	3	2	2	2	2
1	1	3	2	1	3
…	…	…	…	…	…
0.5	2	1	1	2	4
0.5	3	2	2	3	2

4.2 Decision Tree Construction Processing

The decision tree is constructed based on learning strategy model in Sect. 2.2 and data obtained from the questionnaires. In the learning strategy decision tree, we could get a best suitable learning strategy in the decision tree for user under current condition. In the final learning strategy recommendation list, we provide a learning strategy set which contains the best suitable learning strategy, and other alternative learning strategies. Learners can choose one that he/she regards it best for him/her.

Fig. 5 is the decision tree we constructed by decision tree construction algorithm C4.5 described in Sect. 3.1. The results present learning strategy rules which are according to the reality situation in the learning procedure. As shown in Fig. 5, a leaner prefers to choose affective strategy to do low level test when his/her test results and mood is poor and he/she is a learner who don't choose watching video as his/her learning way. Learner prefers to choose repeat strategy when his/her test results and mood is poor and he/she is a learner who would like to choose watching video as his/her learning way.

The quantity of test records is *2%* of whole data records. The quantity of raining records is *98%* of whole data records. We choose records as training records or test records randomly. The average support value of the rules in the decision tree is *51.8%* and the average confidence value of the rules in the decision tree is *26.7%*. To some extent, these rules of choosing leaning strategies described in Fig. 5 reflect the actual learners' learning behaviors. We could adopt decision tree construction algorithm C4.5 in our learning strategy recommendation system (BIT-LSS).

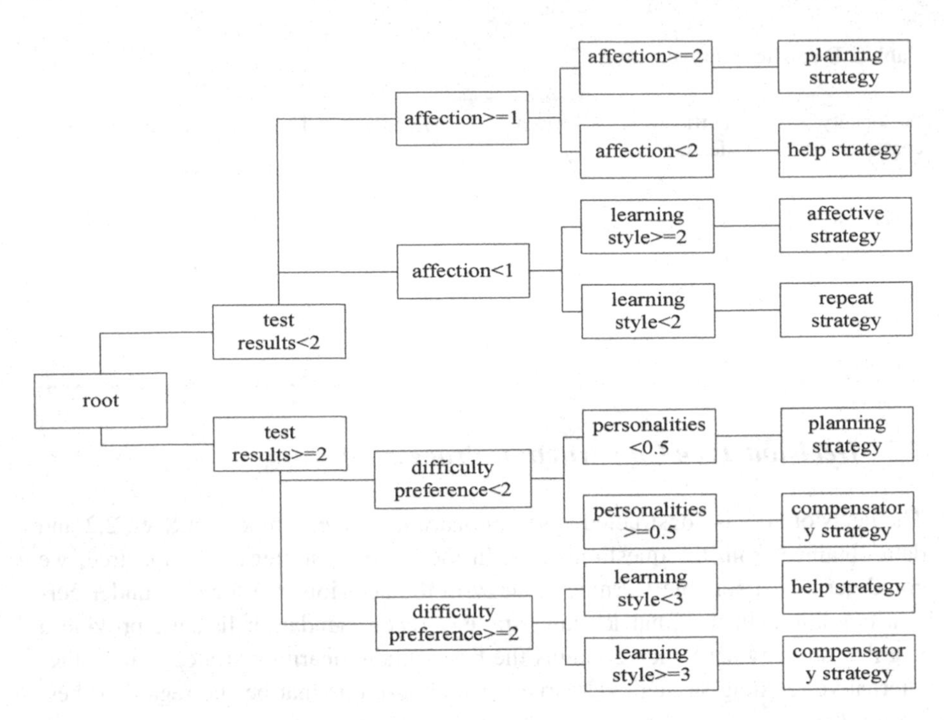

Fig.5 The generated learning strategy decision tree

4.3 Strategy Recommendation Agent

The rules that we choose to recommend strategies in the final list for learners are selected by strategy recommendation controller in strategy recommendation agent. If a learner attends to learn in BIT-LSS for the first time, agent recommends strategies according to the most commonly used learning strategy set. If a learner has learning history, agent recommends strategies just according to the decision tree. If a learner has strategy choosing history, we calculate weight values of strategies according to *(3.4)* and *(3.5)*. The learning strategy recommendation list for Wang who is a learner in BIT-LSS is as Fig. 6 shows. He prefers to choose planning strategy when his test result is poor and his mood is high. We calculate weight values of strategies according to *(3.4)* and *(3.5)* in the candidate strategy set which contains strategy recommended by the strategy decision tree and strategies in adopted history, and the final recommendation list that we get for Wang is planning strategy, social strategy, repeat strategy, compensatory strategy and affective strategy. Wang gives his feedback for the learning strategy recommendation list by giving them different marks. As shown in Fig. 7, Wang marks *2* to planning strategy, *1* to social strategy, *3* to repeat strategy, *2* to compensatory strategy and *1* to affective strategy. We use highest-scoring learning strategy as most popular learning strategy in BIT-LSS.

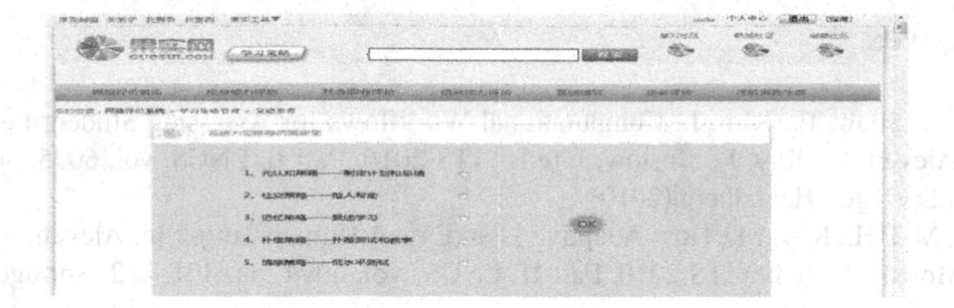

Fig. 6 Learning strategy recommendation list for Wang

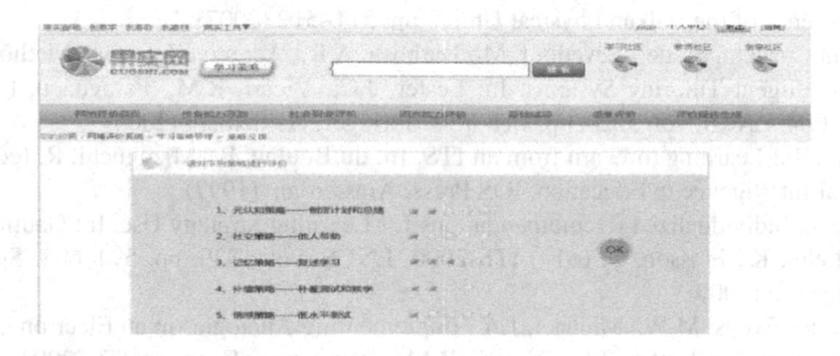

Fig. 7 Wang's feedback for learning strategy recommendation list

5 Conclusions

In this paper, three models of learners, learning strategies and learning strategy-oriented services are proposed. Combining with investigation questionnaires, the C4.5 decision tree construction algorithm is adopted to construct a learning strategy tree. A learning strategy recommendation agent is constructed to help learners make good choices in learning strategy selection in BIT-LSS. In the experiments, the generated learning strategy tree is proven to be effective. The learning strategy recommendation agent recommends strategy lists for learners. It supports BIT-LSS to provide learning strategy recommendation list for novice, intermediate and advanced learners.

In our future work, we will expand the scope of questionnaire to optimize decision attributes. We will consider making adjustments for general learning strategy group with real applications.

Acknowledgments. This work was supported by the National Basic Research Program of China (973 Program No 2012CB720700)", CETV learning mall project, Key Foundation Research Projects of Beijing Institute of Technology(grant no.3070012231001), Beijing Municipal Commission of Education (grant no.1320037010601), the 111 Project of Beijing Institute of Technology and the National Natural Science Foundation of China (project no. 61003263).

References

1. Ma, J., Shaw, E., Kim, J.: Computational Workflows for Assessing Student Learning. In: Aleven, V., Kay, J., Mostow, J. (eds.) ITS 2010, Part II. LNCS, vol. 6095, pp. 188–197. Springer, Heidelberg (2010)
2. Chi, M.T.H., Roy, M.: How Adaptive Is an Expert Human Tutor? In: Aleven, V., Kay, J., Mostow, J. (eds.) ITS 2010, Part II. LNCS, vol. 6094, pp. 401–412. Springer, Heidelberg (2010)
3. Selçuk, G.S., Çalişkan, S., Erol, M.: Learning Strategies of Physics Teacher Candidates: Relationships with Physics Achievement and Class Level. In: Sixth International Conference of the Balkan Physical Union, pp. 511–512 (2007)
4. Glória Curilem, S., de Azevedo, F.M., Barbosa, A.R.: Adaptive Interface Methodology for Intelligent Tutoring Systems. In: Lester, J.C., Vicari, R.M., Paraguaçu, F. (eds.) ITS 2004. LNCS, vol. 3220, pp. 741–750. Springer, Heidelberg (2004)
5. Linton, F.: Learning to Learn from an ITS. In: du Boulay, B., Mizoguchi, R. (eds.) Artificial Intelligence in Education. IOS Press, Amsterdam (1997)
6. Bull, S.: Individualized Recommendations for Learning Strategy Use. In: Gauthier, G., VanLehn, K., Frasson, C. (eds.) ITS 2000. LNCS, vol. 1839, pp. 594–603. Springer, Heidelberg (2000)
7. Lulis, E., Evens, M.W., Michael, J.A.: Implementing Analogies in an Electronic Tutoring System. In: Lester, J.C., Vicari, R.M., Paraguaçu, F. (eds.) ITS 2004. LNCS, vol. 3220, pp. 751–761. Springer, Heidelberg (2004)
8. Hashemi, M.: The Impact of Gender on Language Learning Strategies of Iranian Learners. International Journal of Academic Research, Part I, 280–285 (2011)
9. Arroyo, I., Woolf, B.P., Royer, J.M., Tai, M., English, S.: Improving Math Learning through Intelligent Tutoring and Basic Skills Training. In: Aleven, V., Kay, J., Mostow, J. (eds.) ITS 2010, Part I. LNCS, vol. 6094, pp. 423–432. Springer, Heidelberg (2010)
10. Rubin, J.: What the 'Good Language Learner' Can Teach Us. TESOL Quarterly 9(1) (1975)
11. Zhang, W.S.: Research on Learning Strategy-Oriented Service model in E-Learning. Beijing Institute of Technology (2012)
12. Scarcella, R., Oxford, R.: The Tapestry of Language Learning: The Individual in the Communicative Classroom, p. 63. Heinle & Heinle, Boston (1992)
13. Oxford, R.L.: Language Learning Styles and Strategies: an Overview. In: GALA 2003, pp. 1–25 (2003)
14. Hu, J.L., Deng, J.B., Sui, M.X.: A new approach for decision tree based on principal component analysis. In: 2009 International Conference on Computational Intelligence and Software Engineering, pp. 1–4 (2009)
15. Quinlan, J.R.: C4.5: Programs for machine learning. Morgan Kauffman publishers (1993)

Author Index